IN THE BEGINNING

and Other Essays on Intelligent Design

IN THE BEGINNING
and Other Essays on Intelligent Design

GRANVILLE SEWELL

Mathematics Department
University of Texas at El Paso
El Paso, Texas 79968

Discovery Institute Press
Seattle, Washington

Description

In this wide-ranging collection of essays on origins, mathematician Granville Sewell looks at the big bang, the fine-tuning of the laws of physics, and the evolution of life. He concludes that while there is much in the history of life that seems to suggest natural causes, there is nothing to support Charles Darwin's idea that natural selection of random variations can explain major evolutionary advances. Sewell explains why evolution is a fundamentally different and much more difficult problem than others solved by science, and why increasing numbers of scientists are now recognizing what has long been obvious to the layman, that there is no explanation possible without design. This book summarizes many of the traditional arguments for intelligent design, but presents some powerful new arguments as well.

Granville Sewell is Professor of Mathematics at the University of Texas at El Paso. He has written three books on numerical analysis, and is the author of a widely-used finite element computer program.

Copyright Notice

Publisher's Note

This book is part of a series published by the Center for Science & Culture at Discovery Institute in Seattle. Previous books include *Why Is a Fly Not a Horse?* by Guiseppe Sermonti, *The Deniable Darwin and Other Essays* by David Berlinski, *Darwin's Conservatives: The Misguided Quest*, by John G. West, and *Traipsing into Evolution: Intelligent Design and the Kitzmiller vs. Dover Decision*, by David K. DeWolf et al.

Library Cataloging Data

In the Beginning and Other Essays on Intelligent Design by Granville Sewell (1948-)
148 pages, 6 x 9, 229 x 152 mm
Library of Congress Control Number: 2009938790
BISAC Subject: SCI027000 SCIENCE / Life Sciences / Evolution
BISAC Subject: SCI015000 SCIENCE / Cosmology
BISAC Subject: SCI055000 SCIENCE / Physics
BISAC Subject: SCI080000 SCIENCE / Essays
BISAC Subject: MAT039000 MATHEMATICS / Essays
ISBN-13: 978-0-9790141-4-7 ISBN-10: 0-9790141-4-X (paperback)

Publisher Information

Discovery Institute Press, 208 Columbia Street, Seattle, WA 98104
Internet: http://www.discovery.org/
Published in the United States of America on acid-free paper.
First Edition, First Printing. January 2010.

Contents

Preface

This book is a collection of short articles on intelligent design and related topics. Several of the articles have been previously published, some are unpublished. A summary of the essays follows:

1. *In the Beginning.* This unpublished essay presents some of the evidence for the astonishing but now widely-accepted idea that the universe had a beginning, in a "big bang" about 15 billion years ago. Since there were no natural causes before Nature came into existence, all theories on origins now involve speculation as to the nature of the supernatural forces—intelligent or unintelligent—that brought our universe into existence.

2. *Design in the Laws of Nature.* This unpublished essay discusses some of the fortunate but improbable features of our universe which were required for the development of life. The "fine-tuning" of the fundamental constants of physics, and of the initial conditions of our universe, is based on widely-accepted and published scientific research, and has forced atheists to hypothesize the existence of many universes, with different constants and conditions,

to avoid the obvious explanation of design. The author notes that not only are the basic constants of physics fine-tuned, but so is the fundamental equation itself which underlies all of chemistry, the Schrödinger equation.

3. *A Mathematician's View of Evolution.* This essay was published in *The Mathematical Intelligencer* in 2000. In discussing Michael Behe's "irreducible complexity" arguments, the author draws an analogy between the development of life, as it appears in the fossil record ("most taxa appear abruptly... gaps among known orders, classes and phyla are systematic and almost always large"), and the 20-year "evolution" of his partial differential equation solving software. This software also evolved through the release of many new versions, each with obvious similarities to previous versions, but also with large gaps where major new features appeared and smaller gaps where minor new features appeared. Major, complex, evolutionary advances, involving new features, require the addition of many interrelated and interdependent pieces. Like major improvements to computer programs, they are not reducible to chains of tiny improvements.

4. *Postscript in 1985 Book.* This is an appendix to the author's 1985 Springer-Verlag book, *Analysis of a Finite Element Method: PDE/PROTRAN*, where the analogy between the evolution of life and the evolution of this software project is first discussed. The content is similar to the previous chapter, but is interesting in that it anticipates some of the current arguments for intelligent design (ID), 11 years before publication of Behe's now-classic work, *Dar-*

win's Black Box [Behe 1996].

5. *Can 'Anything' Happen in an Open System?* This is based on a 2005 online article in the *American Spectator*. The origin and development of life seem to violate the second law of thermodynamics in a clear and spectacular way; however, such arguments are routinely dismissed by saying that the second law does not apply to open systems, such as the Earth. The author counters this idea with the tautology that "if an increase in order is extremely improbable when a system is closed, it is still extremely improbable when the system is open, unless something is entering which makes it *not* extremely improbable." In an appendix to a 2005 John Wiley mathematics text, reproduced in this chapter, the author looks at the usual equations for the second law as it applies to heat conduction and diffusion, and shows that they actually confirm this common sense interpretation, rather than the idea that *anything* can happen in an open system. The conclusion: "If we found evidence that DNA, auto parts, computer chips, and books entered through the Earth's atmosphere at some time in the past, then perhaps the appearance of humans, cars, computers, and encyclopedias on a previously barren planet could be explained without postulating a violation of the second law here. But if all we see entering is radiation and meteorite fragments, it seems clear that what is entering through the boundary cannot explain the increase in order observed here."

6. *My Failed Simulation.* This essay was published online in 2008 in *Human Events*. The strongest argument for ID is to clearly state the alternative

view, which is that physics explains all of chemistry (probably true), chemistry explains all of biology, and biology completely explains the human mind; thus, physics alone explains the human mind. This thought experiment is designed to help those who dismiss ID as unscientific, to think about what it is they really believe.

7. *How Evolution Will Be Taught Someday.* Also published online in 2008 in *Human Events*, this essay is a short overview of what ID is all about. It was designed to present the basic ideas and arguments for ID in a very short summary form, for people who don't want to spend more than five minutes on the topic. One of the main points is that all of the "overwhelming evidence" for evolution is really of the form "this just looks like natural causes," and "we have found natural explanations for so many other phenomena," there is nothing to support the idea that natural selection can explain the complexity of life. In this essay, the author predicts that future biology texts will refer to evolution as a mysterious "natural" process, which scientists do not now understand, but still hope to understand some day.

8. *The Supernatural Element in Nature.* This unpublished essay looks at the history and philosophical consequences of quantum mechanics, which has blurred the distinction between what is natural and what is supernatural. When we try to reduce all of reality to matter in motion, we find quite a surprise: there at the bottom, controlling the motion of matter, is the remarkable Schrödinger equation of quantum mechanics, which tells us that science is an

entertaining and useful tool to help us understand our world, but it does not have all the answers, and never will.

9. *The Scientific Theory of Intelligent Design.* This unpublished essay looks at some of the issues raised in the debate as to whether or not intelligent design is really "science." While much of the material in this chapter will be familiar to those acquainted with ID, there is some relatively new information here also, for example, a discussion of the "front-loading" being discovered in primitive animals, which is completely fatal for Darwinism.

E. *Epilogue: Is God Really Good?* This essay, published in the Indian journal *AntiMatters* in 2008, looks at one of the most powerful theological arguments *against* ID, the problem of pain. While it may seem out of place in an otherwise scientific book, it is not an unrelated foray into theology, but is relevant to the rest of the book.

About the Author

Granville Sewell is Professor of Mathematics at the University of Texas at El Paso (UTEP). He completed his PhD in Mathematics at Purdue University, and has subsequently been employed by (in chronological order) Universidad Simon Bolivar (Caracas), Oak Ridge National Laboratory, Purdue University, IMSL Inc. (Houston), UTEP, The University of Texas Center for High Performance Computing (Austin), and Texas A&M University, and is currently back at UTEP. He spent one semester (Fall 1999) teaching at Universidad Nacional de Tucuman in Argentina, on a Fullbright grant, and returned to Universidad Simon Bolivar to teach summer courses in 2005 and 2008. Dr. Sewell has written three books on numerical analysis, and is the author of a widely-used finite element computer program (video at www.vni.com/pde2d).

1

In the Beginning

1.1 The Expanding Universe

Using the Doppler shift to measure the speeds of distant stars, astronomer Edwin Hubble discovered in 1929 that all but a few of the closest galaxies are moving away from us, and that the speed at which each is moving away is approximately proportional to its distance from us. The proportionality constant is called the Hubble constant, so that the speed with which a galaxy recedes from us is approximately given by H times its current distance from us. Current estimates of the Hubble constant are in the neighborhood of $H=7 * 10^{-11}$/year. In other words, a galaxy which is currently R miles away from us is now receding from us at the rate of about H*R miles/year. At this rate, every 1/H years it recedes H*R*(1/H) = R miles, and so if we assume that it has always been receding from us at this same velocity, it is easy to see that 1/H years ago it must have been 0 miles from us! Since the same thing can be said for any distant galaxy, we calculate that all the galaxies would have been very closely clustered together some $1/H \approx 15$ billion years ago, when the current expansion must have begun. Actually, the gravitational attraction between stars and galaxies must be gradually slowing down this expansion, so that the expansion rate in the past must have been even higher. This means that 15 billion years is an upper limit to the time which has passed since the beginning of the expansion of the universe. In fact, if the only significant force between astronomical bodies, the force of gravity, works against expansion, what force sent the galaxies hurdling away from each other through space 15 billion years ago? It must have required quite an explosion, quite a "big bang," to overcome the force of gravity and send all this cosmic debris flying out through space.

In order to use the apparent expansion of the universe to interpret either the past or the future, we must make some assumptions. The basic assumption of cosmology is the "cosmological principle," which states that the universe appears essentially the same (i.e., modulo minor local variations) to observers at any point in it. This means that matter must be distributed more or less evenly throughout the universe, so that we may treat the matter in it as if it were a gas of more or less constant density. Of course there are local variations: an observer in the middle of the Milky Way galaxy will see a lot more stars in his sky than an observer halfway between galaxies, but we are talking about the larger view, where galaxies can be considered the "molecules" of the gas!

Notice that the cosmological principle also implies that the universe cannot have boundaries, because then it would appear different to an observer near the edge than to one far from the boundary (a boundary would be rather problematic in any case!). Surprisingly, this does not necessarily mean that the universe has to be infinite in volume, with an infinite quantity of matter distributed throughout. According to Einstein's general theory of relativity, the universe could be "finite but unbounded." A "finite but unbounded" universe is a concept which can only be grasped intuitively by saying that the universe would appear to us like the surface of a sphere would appear to a creature who can only imagine two dimensions. The surface of a sphere has a finite area but it has no boundaries, and it looks the same to all its two dimensional inhabitants, wherever they live on it. Our universe is three dimensional but, according to Einstein, we can think of it as being embedded in a higher dimensional space.

According to the general theory of relativity, space is warped, or curved, in the vicinity of matter, and if the density of matter (which we are still assuming to be approximately the same throughout the universe) is greater than some critical density, the curvature of space will be large enough to ensure that the universe is closed, comparable to the surface of a sphere. If the density is less than this critical density, we can compare our universe to a curved but open surface in 3D space, such as a hyperbolic paraboloid (Figure 1.1).

Is the cosmological principle justified by the evidence? Although the part of the universe that we can see does appear extremely homogeneous, the justification for the cosmological principle is really as much philosophical as observational, since we can presumably see only a small portion of the whole universe. However, it is hard to see how we could assume anything else—if the cosmological assumption is not valid, there is not much hope for modeling the expansion of the universe.

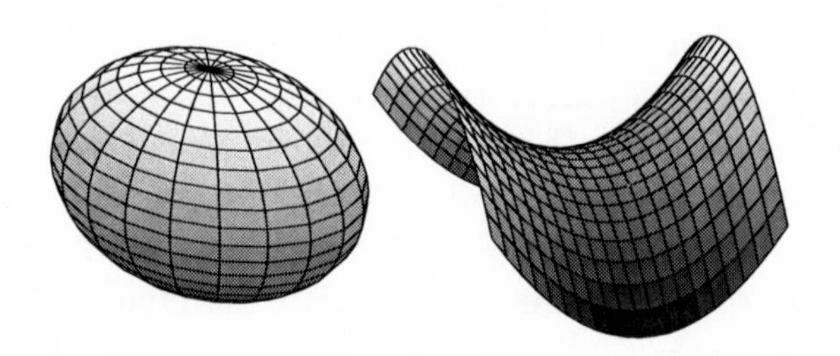

Figure 1.1. Closed and Open 2D Universes

Now we are ready to talk about the future of the universe. Will the universe continue to expand forever, or will the attractive force of gravity slow the expansion rate to zero, then cause the universe to begin contracting? To answer this question would seem to require the more modern ideas about gravity given by Einstein's general theory of relativity (which are far beyond the scope of this book), but in fact it is possible to pose this question in a mathematically correct way using the classical (Newtonian) theory of gravity. This is done in the Supplement (Section 1.5).

1.2 The Big Bang

In the Supplement, an equation for $r(t)$, the normalized size of the universe (normalized so that $r = 1$ today) is derived and it is shown that there is a critical value ρ_c of the density of matter in the universe such that if the current density ρ_0 is larger than this value, the gravitational attraction between galaxies will be strong enough to eventually stop the expansion of the universe and cause it to contract; otherwise it will continue expanding forever. It turns out that ρ_c is also exactly the critical density which will "close" space. In other words, if the current density is larger than ρ_c not only can we conclude that the expansion of the universe will eventually halt, but also that the universe is finite but unbounded (closed). Thus a closed universe will eventually contract, while an open universe will continue expanding forever.

Is ρ_0 larger than the critical value? The best current estimates say no, so that the universe is infinite in size, and will continue expanding forever. But it is very difficult to estimate the density of matter in the universe, and the current estimates of ρ_0 are close enough to ρ_c to leave considerable doubt as to the size and future of the universe. I believe the universe is finite, because, in my opinion, "infinity" is an abstract concept which exists only in pure mathematics, I don't believe there could really be an infinite amount of anything. But I could be wrong, since it is a surprising universe we live in.

If we give a rock an initial upward velocity which equals or exceeds the Earth's escape velocity, it will continue upward forever and never return to Earth. If we give it an upward velocity less than this escape velocity, the attraction of the Earth's gravity will cause the rock to slow down and finally stop and begin falling back down. But in any case, if we see a rock flying upward through the air, even if we are unable to calculate whether its future holds a return to Earth or an eternal flight through outer space, we can confidently "predict" its past.

In a similar manner, whatever the future of the universe may be, it is clear that at some time in the past $r(t)$ was very small, and the matter in the universe was very tightly packed. If we solve the equation (1.2), derived in the Supplement (Section 1.5), for $r(t)$ we find that, no matter what value we assign to ρ_0, at a particular negative (past) value of t, the normalized size of the universe was $r = 0$, and the velocity of expansion was $r' = \infty$. The value of t for which $r = 0$ depends on our estimate of ρ_0, but we have already seen that it cannot have been more than 15 billion years ago. For example, if $\rho_0 = \rho_c$, then it is shown in the Supplement that the big bang must have occurred about 10 billion years ago. That $r' = \infty$ when $r = 0$ can be seen directly from (1.2), and it reflects the fact that the gravitational attraction between bodies becomes infinite as their separation approaches zero, and so an infinitely large initial velocity would be required to overcome the infinitely strong attraction of gravity associated with the state $r = 0$. This is suggestive of a very "big bang"!

The justification from Newtonian gravitational theory for the equations derived in the Supplement would seem to break down when r is close to 0, because other forces are no longer negligible then, but the real justification for these equations is not Newtonian physics but the general theory of relativity. A model of the universe which is based on the general theory of relativity still predicts a singularity, with $r = 0$ and $r' = \infty$ in the finite past. The currently accepted model, called the "standard model," still says that the universe arose from nothing with a "big bang."

For a while scientists were divided between the big bang theory and the "steady state" theory of the universe. The steady state theory holds that the average density of the universe is maintained at a constant level as the universe expands, by the creation (somehow) of new matter. However, there are several theoretical and experimental reasons why the steady state model has now been rejected in favor of the big bang theory. In particular, the 1965 discovery by a pair of radio astronomers of a background of microwave radiation permeating the universe was spectacular confirmation of one of the predictions of the big bang model. Proponents of the big bang theory had calculated that at a time shortly after its beginning, when the temperature of the universe was about $3,000^o$ Kelvin, the universe must have been filled with high energy, that is, short wavelength, photons. They calculated that the entire universe at that point would function as a "blackbody," and that the photon wavelengths would be distributed in the manner characteristic of blackbody spectra. As the universe expanded, this radiation cooled and by now, they calculated, it should have an equivalent temperature of about 3° Kelvin. With the drop in temperature, the photon wavelength distribution would shift and would now have a peak in the microwave range.

It was this 3^oK remnant microwave radiation, emanating not from any particular astronomical object but from the entire sky, that Bell Telephone Laboratory radio astronomers Arno Penzias and Robert Wilson discovered in 1965. Not only was the peak close to the expected value, but the form of the observed wavelength distribution curve conformed closely to the predicted shape. Robert Jastrow, founder and director of NASA's Goddard Institute for Space Studies, and professor at Columbia University, describes the discovery of the background microwave radiation in layman's language [Jastrow 1978]:

> The two physicists were puzzled by their discovery. They were not thinking about the origin of the universe, and they did not realize

> that they had stumbled upon the answer to one of the cosmic mysteries. Scientists who believed in the theory of the big bang had long asserted that the universe must have resembled a white-hot fireball in the first moments after the big bang occurred. Gradually, as the universe expanded and cooled, the fireball would have become less brilliant, but its radiation would never have disappeared entirely. It was the diffuse glow of this ancient radiation, dating back to the birth of the universe, that Penzias and Wilson apparently discovered.
>
> No explanation other than the big bang has been found for the fireball radiation. The clincher, which has convinced almost the last doubting Thomas, is that the radiation discovered by Penzias and Wilson has exactly the pattern of wavelengths expected for the light and heat produced in a great explosion. Supporters of the steady state theory have tried desperately to find an alternative explanation, but they have failed. At the present time, the big bang theory has no competitors.

We suggested earlier that our 3D universe, if finite, may appear to us like the surface of a sphere would appear to a 2D creature who cannot even comprehend the concept of a third dimension. The inflation of a sphere or balloon is perhaps a better analogy than an explosion to illustrate the expanding universe. If air is pumped into a balloon, it will expand in such a way that every point in this 2D universe (the balloon surface) recedes from every other point. This analogy also helps us understand how the attraction of gravity could cause the expansion to slow, or to reverse itself, even though the pull of gravity on any star is the same in every direction. The word "explosion" implies that a volume of empty space is already there, and an explosion at one point in that volume sends debris flying out in all directions through the already-existing space. But our expanding universe is more like the surface of a sphere whose radius has expanded from $r = 0$ to its current size. There was no universe, not even an empty one, before the big bang, and it is the entire universe—empty space and all—which is expanding. $r = 0$ does not mean a very small, dense, universe, it means nothing existed: not only no matter or energy, but no space or time either!

1.3 The Finite Age of the Universe

Although the discovery of the background microwave radiation permeating the universe is one reason that the big bang theory is the "standard" model today, there are other reasons to believe the universe had a beginning, and most are consistent with the time frame of 15 billion years estimated from the expansion rate of the universe. The ages of various celestial objects can been estimated, and all are found to be less than 15 billion years old. For example, radioactive dating techniques can be used to compute the age of a meteorite; the fraction of a radioactive isotope remaining tells us how many half-lives have passed since the isotope was created.

The fraction of the matter in the universe represented by hydrogen is continually decreasing, as hydrogen atoms in the stars fuse to make helium and other heavier elements. If the universe were infinitely old, all the hydrogen would presumably have been consumed long ago.

A similar argument is based on the second law of thermodynamics, which states that the "entropy" (disorder) in the universe continually increases. Every time hot and cold substances exchange heat, or two gases mix, or mechanical energy is converted by friction into heat energy, the entropy of the universe increases irreversibly—the universe "winds down" a little. But, again, if the universe were infinitely old, the continual increase in randomness predicted by the second law of thermodynamics would ensure that all that would have been left by now would be a homogeneous, "wound down" universe.

Russian astrophysicists Y.B. Zel'Dovich and I.D. Novikov [Zel'Dovich and Novikov 1983] argue that even if we conjecture that the universe goes through cycles of expansion and contraction (and they see no evidence of any repulsive force which could turn contraction back into expansion) the second law of thermodynamics still guarantees that the age of the universe is finite. They write: "It follows from this that the universe has lived through only a finite number of cycles in the past and has a finite time of existence because in each cycle the entropy increases by a finite amount. For an infinite number of cycles, therefore, the specific entropy would be infinite;[1] but this is not the case."

It is conceivable, though it seems extremely unlikely, that evidence will someday be uncovered which forces us back to a steady-state or oscillating universe theory. But it is inconceivable that natural processes will be discovered which can reverse the normal flow of entropy, and cause disorder to reorganize itself into order. Thus Nature's irreversible tendency toward disorder will not allow us to avoid the problem of a true beginning of time, of a

1. The authors should have said, "maximal."

"moment with no moment preceding it" (as Arthur Eddington put it).

1.4 Philosophical Implications

In the introduction to his book *The First Three Minutes* [Weinberg 1977] Steven Weinberg wrote:

> How then did we come to the 'standard model'? And how has it supplanted other theories, like the steady state model? It is a tribute to the essential objectivity of modern astrophysics that this consensus has been brought about, not by shifts in philosophical preference or by the influence of astrophysical mandarins, but by the pressure of empirical data.

To say that rejection of the steady state model in favor of the big bang theory was not due to shifts in philosophical preference is an understatement, because many scientists would agree with Weinberg that the steady state model is "philosophically far more attractive." Einstein introduced an arbitrary additional term into his equations of general relativity in an attempt (which he later regretted) to avoid the expanding universe solution. Robert Jastrow [Jastrow 1978] writes that:

> Some prominent scientists began to feel the same irritation over the expanding universe that Einstein had expressed earlier. Eddington wrote in 1931, 'I have no ax to grind in this discussion, but the notion of a beginning is repugnant to me. The expanding universe is preposterous... incredible, it leaves me cold.'

Carl von Weizsäcker [von Weizsäcker 1964] recounts the reaction of German chemist Walther Nernst to the discovery that time had a beginning:

> He said, the view that there might be an age of the universe was not science. At first I did not understand him. He explained that the infinite duration of time was a basic element of all scientific thought, and to deny this would mean to betray the very foundations of science. I was quite surprised by this idea and I ventured the objection that it was scientific to form hypotheses according to the hints given by experience, and that the idea of an age of the universe was such a hypothesis. He retorted that we could not form a scientific hypothesis which contradicted the very foundations of science. He was just angry, and thus the discussion, which was continued in his private library, could not lead to any result. What impressed me about Nernst was not his arguments; what impressed me was his anger. Why was he angry?

The reason that many scientists were reluctant to accept the big bang is obvious: it points out the incompleteness of science. If the goal of science is, as Joseph Le Conte [Le Conte 1888] put it, to explain how "each state or condition grew naturally out of the immediately preceding," then this pursuit meets a dead end in the big bang, for the chain of causality must end with the beginning of time. The implications of the discovery that the entire universe—matter, energy, space and time—had a true beginning are enormous, and do not yet seem to have sunk in to our scientific consciousness; many scientists still gloss over the big bang as if it were just another explosion. Many scientists still tend to think of religions as systems of beliefs which have no root in science, and of atheism as the absence of any such unprovable beliefs. The truth is that now all theories of origins, theistic or atheistic, involve speculation as to the nature of the supernatural force which created our universe out of nothingness, because there were no "natural" causes before Nature came into existence. The question is only, was it an intelligent or an unintelligent supernatural force that created time, space, matter and energy out of nothingness?

Some religious people do not like the big bang theory because they believe it is an attempt to explain scientifically how our universe came into existence. But while the big bang theory attempts to explain what happened from the very early stages onward, it does not attempt to explain how or why our universe came into being from nothingness—how could any scientific theory ever do that? It only states that, according to the evidence, that is exactly what happened.

British physicist Edmund Wittaker [quoted in Jastrow 1978] stated what other scientists had to be thinking: "What came before the beginning? There is no ground for supposing that matter and energy existed before and were suddenly galvanized into action. For what could distinguish that moment from all other moments in eternity? It is simpler to postulate creation *ex nihilio*—Divine will constituting Nature from nothingness."

1.5 Supplement: A Model for the Expanding Universe

If I dig a hole to the center of the Earth and excavate a small chamber there, I will be able to float about weightlessly in my chamber, because the gravitational attraction of the Earth on my body is equally strong in all directions. What if I only tunnel halfway to the center? As you might expect, I will weigh less than I did on the surface, but I will not be completely weightless; there is still a net force toward the center of the Earth. It can be shown by summing up the gravitational forces exerted on my body by all the molecules in the Earth (thanks to integral calculus, this is not as difficult as it sounds), that the net force exerted by that portion of the Earth which is closer to the surface than I am is zero, so that only the portion closer to the center than I contributes to my weight. If I could hollow out the inner core of the Earth, the entire hollow core would be a giant weightless chamber and I could float about in it, because at any point within the hollow core the tug of gravity would be the same in every direction. Thus the Earth's gravitational attraction on my body can be calculated by throwing away the outer shell and pretending that I am standing on the surface of a smaller planet, half the diameter of the Earth.

Now let us replace the Earth in our story by the entire universe, and let us take our planet to be the center of the universe. (Copernicus showed us that the Earth is not the center of the universe, but the cosmological principle and the theory of relativity tell us that it is as good a center as any!) Then consider a certain galaxy (A) whose distance from the Earth is given as a function of time by

$R(t)$, and let us calculate the "weight" of that galaxy—that is, the gravitational force with which the rest of the matter in the universe pulls it toward the center (us). By the same reasoning as used above, we conclude that we can ignore all matter further away from the center than A, and calculate the force pulling A toward the center as the force of gravity between A and a sphere of matter whose center is the Earth and whose radius is R. According to Newtonian gravitational theory, this force is $GMm/R(t)^2$, where G is the universal gravitation constant, m is the mass of A, and M is the mass of the above described sphere, on whose surface A rests. M is just the density $\rho(t)$ times the volume of this sphere, $\frac{4}{3}\pi R(t)^3$. As the universe expands or contracts, this sphere will expand or contract proportionally, so the quantity of matter in the sphere will remain constant. Then we may use the current ($t = 0$) values for $\rho(t)$ and $R(t)$, and write M = $\frac{4}{3}\pi\rho_0 R_0^3$.

This gives, for the gravitational force on A:

$$GMm/R(t)^2 = \frac{4}{3}\pi\rho_0 R_0^3 Gm/R(t)^2$$

By Newton's second law, then, the acceleration (R'') of A is equal to this force divided by the mass (m) of A:

$$R''(t) = -\frac{4}{3}\pi\rho_0 R_0^3 G/R(t)^2$$

where the negative sign is used because the acceleration is negative, that is, gravity decelerates (slows down) the expansion.

The initial conditions for this differential equation are obtained by noting that at $t = 0$ (now) we have $R(0) = R_0$ and $R'(0) = HR_0$, since the rate at which any galaxy is receding from us is supposed to be approximately H (the current Hubble constant) times its distance from us.

If $r(t)$ is defined to be $R(t)/R_0$, r can be interpreted as the size of the universe normalized to make the current size equal to 1. Then the differential equation and initial conditions simplify to:

$$\begin{aligned} r'' &= -\frac{4}{3}\pi G\rho_0/r^2 \qquad (1.1) \\ r(0) &= 1 \\ r'(0) &= H \end{aligned}$$

Now there is an objection which the reader may raise to the way in which (1.1) was derived. It may be argued that there is no net gravitational force on either the Earth or A, because in either case the pull of the rest of the matter in the universe is equally strong in all directions—either can be considered the center of the universe. The answer is that the ultimate justification for (1.1) comes from the general theory of relativity. However, if we look at any sphere of small (cosmologically speaking!) radius, the general theory of relativity allows us to ignore the gravitational effects of material outside that sphere, and to use Newtonian gravitational theory inside the sphere. And we will get equation (1.1) using classical ideas if we take the universe to be a sphere of arbitrary radius, with center at any particular point—Earth, galaxy A, or a neutral third party.

Using techniques found in any elementary differential equations text we find that (1.1) implies:

$$r' = \sqrt{(8\pi/3)G\rho_0/r + C} \qquad (1.2)$$

where C is found, by applying the initial conditions, to be

$$C = H^2 - \frac{8}{3}\pi G\rho_0$$

Now if C is positive, that is, if $\rho_0 < 3H^2/(8\pi G) \equiv \rho_c$, then it is clear from (1.2) that the universe will continue expanding forever, since r' will always be positive. On the other hand, if $\rho_0 > \rho_c$, C will be negative and there will be a value of r which will make $r' = 0$, so that when the normalized size of the universe reaches that value of r, it will stop expanding and begin to contract. This contraction would presumably continue until the universe ends in a "big squeeze."

The differential equation (1.2) can be further solved for $r(t)$ using standard differential equations techniques, but the resulting solution is rather complicated to write out. However, for the case $\rho_0 = \rho_c$, which is thought to be reasonably close to correct, C=0 and (1.2) reduces to $r' = \frac{H}{\sqrt{r}}$ which can be easily solved (remembering that $r(0) = 1$) to give $r(t) = (\frac{3}{2}Ht + 1)^{2/3}$. In this case, we can see that $r = 0$ and $r' = \infty$ when $t = -\frac{2}{3}H^{-1}$, which places the big bang about 10 billion years ago. For other values of ρ_0, the solution is more complicated, but still predicts that $r = 0$ in the finite past, less than $H^{-1} \approx 15$ billion years ago. Some estimates of ρ_0 are around $0.01\rho_c$; that would make the age of the universe about $0.98H^{-1}$.

2

Design in the Laws of Nature

2.1 The 'Fine-Tuning' of the Laws of Physics

The development of models of the early universe involves primarily theoretical calculations, intended to reconstruct what must have happened in the aftermath of the big bang. These computations, and many others made by physicists, show that we are the beneficiaries of some very lucky coincidences. In an interview published in [Varghese 1984], Columbia University astronomer Robert Jastrow discusses what he calls "the most theistic result ever to come out of science":

> According to the picture of the evolution of the universe developed by the astronomer and his fellow scientists, the smallest change in any of the circumstances of the natural world, such as the relative strengths of the forces of Nature, or the properties of the elementary particles, would have led to a universe in which there

could be no life and no man.

As an example, Jastrow cites the forces binding the nuclei of atoms together. If the nuclear force were increased in strength by a small amount, he says, this attraction would have been sufficient to cause all hydrogen nuclei (protons) to fuse together into helium during the early stages of the universe, and there would be no hydrogen left to fuel the stars. On the other hand, if the nuclear force were slightly decreased in strength, the attraction would have been insufficient to drive the nuclear fusion reactions which created elements heavier than helium (such as carbon and oxygen), and it is impossible to imagine how any complex life forms could be constructed out of hydrogen and helium alone.

Jastrow continues:

> It is possible to make the same argument about changes in the strengths of the electromagnetic force, the force of gravity, or any other constants of the material universe, and so come to the conclusion that in a slightly changed universe there could be no life, and no man. Thus according to the physicist and the astronomer, it appears that the universe was constructed within very narrow limits, in such a way that man could dwell in it. This result is called the anthropic principle.
>
> Some scientists suggest, in an effort to avoid a theistic or teleological implication in their findings, that there must be an infinite number of universes, representing all possible combinations of basic forces and conditions, and that our universe is one of an infinitely small fraction, in this great plenitude of universes, in

which life exists.

Now the Darwinist might argue that a different universe, which might be hostile to life as we know it, would only have resulted in life forms which are adapted to different conditions. However, we are not talking about conditions which are hostile to life as we know it on Earth, but rather conditions so hostile that any imaginable form of life would be impossible. In *The Problems of Physics*, A.J. Leggett [Leggett 1987] lists several ways in which the development of life depends sensitively on the values of the universal constants, and says,

> The list could be multiplied endlessly, and it is easy to draw the conclusion that for any kind of conscious beings to exist at all, the basic constants of Nature have to be exactly what they are, or at least extremely close to it. The anthropic principle then turns this statement around and says, in effect, that the reason the fundamental constants have the values they do is because otherwise we would not be here to wonder about them.

Physicist Steven Hawking discusses some of these fundamental constants of Nature and says [Hawking 1988], "The remarkable fact is that the values of these numbers seem to have been very finely adjusted to make possible the development of life."

In *Cosmology*, Edward Harrison [Harrison 1981] mentions some other bad things which would happen if certain constants were tampered with:

> We first notice that alterations in the known values of c [speed of light], h [Planck's constant], and e [electronic charge] cause huge changes

> in the structure of atoms and atomic nuclei. Even when the changes are only slight, most atomic nuclei are unstable and cannot exist.... We also find that slight changes in the values of c, G [gravitational constant], h, e, and the masses of subatomic particles cause huge changes in the structure and evolution of stars. The majority of universes will actually not contain any stars at all, and in the few that do, the stars either are nonluminous or are so luminous that their lifetimes are too short for biological evolution.... Our universe is therefore finely tuned, and we would not exist if the constants of Nature had different values.

But we have to ask ourselves not only, why do the gravitational, nuclear and electromagnetic forces have the strengths that they have, and why do electrons, protons and neutrons have the masses and charges they do, but why are there particles at all, and why are there forces between them? We need to wonder not only why the speed of light is 299,792 km/sec, but why are there photons? We should not only wonder why Planck's constant, which appears in the Schrödinger equation, has such a lucky value, but why are the motions of all particles governed by this partial differential equation?[1] One of the most surprising things about our universe is the beautiful way in which mathematical equations can be

---

1. In an N-particle system whose potential energy is given by V, the probability (per unit volume) of finding particle 1 at (x_1, y_1, z_1), particle 2 at (x_2, y_2, z_2), etc, at time t, is $|u(t, x_1, y_1, z_1, ..., x_N, y_N, z_N)|^2$, where u is the solution to the Schrödinger equation
$$-ih\frac{\partial u}{\partial t} = \sum_{k=1}^{N} \frac{h^2}{2m_k}\left[\frac{\partial^2 u}{\partial x_k^2} + \frac{\partial^2 u}{\partial y_k^2} + \frac{\partial^2 u}{\partial z_k^2}\right] - V(x_1, y_1, z_1, ..., x_N, y_N, z_N)u$$
where h is Planck's constant, m_k is the mass of particle k and i is the complex number $\sqrt{-1}$.

used to elegantly model physical processes. In the case of macroscopic processes, such as diffusion or fluid flow, we can derive the equations from more basic processes, so that in these cases we feel we "understand" why the mathematics fits the physics so nicely. But when we get down to the most fundamental particles and forces, we find they *still* obey an elegant mathematical equation, and we have absolutely no idea why—they just do. There is no conceivable reason why the effect that the fundamental forces have on the fundamental particles should be given by the (complex-valued!!) solution to a wave or eigenvalue partial differential equation, except that it results in elements and chemical compounds with extremely rich and useful chemical properties, and gives partial differential equation software developers like me some very interesting applications to solve. In *Partial Differential Equations* [Strauss 2008], Walter Strauss writes, "Schrödinger's equation is most easily regarded simply as an axiom that leads to the correct physical conclusions, rather than as an equation that can be derived from simpler principles.... In principle, elaborations of it explain the structure of all atoms and molecules and so all of chemistry!"

Are we to assume that in all these other universes there are still space and time, gravity and electromagnetic forces, electrons, protons, neutrons and photons, and the behavior of the particles is still governed by the Schrödinger partial differential equation; but the forces, masses and charges, and Planck's constant and the speed of light have different values, generated by some random number generator? Or perhaps the behavior of particles is governed by random types of partial differential equations in different universes, but there are still many universes in which Schrödinger's equation holds, with random values for Planck's constant. No doubt there were some universes which couldn't produce life because their fundamental equation of chemistry looked just like the Schrödinger equation, but with first derivatives in space where there should be second derivatives, or a second derivative in time where there should be a first derivative, or the complex number i was missing, or the linear Vu term was replaced by a nonlinear term Vu^n, where n is not equal to one.[2]

Scientists modeling the big bang have discovered that a universe capable of supporting life requires not only finely-tuned laws, but also initial conditions which are astronomically improbable. Paul Davies, in *Other Worlds* [Davies 1980], appeals to the anthropic principle no fewer than 10 times to explain benevolent features of our universe. Citing the calculations of various physicists and astronomers, he notes that fine-tuning of various laws is required (e.g., the strengths of the strong and weak nuclear forces must be just right), but also shows that, for example, if the matter in the early universe were distributed a tiny bit more—or less—uniformly, or if the material density were a tiny bit higher—or lower, then the resulting universe would have been very hostile toward the conception and development of any form of life. Davies estimates the odds against one of these coincidences to be $10^{1000000000000000000000000000000}$ to 1. And he adds that "there are probably many more features of the

2. Any of the changes listed—and many others not listed—would fundamentally alter the nature of the solutions, and chemistry as we know it would not exist. For example, without the $-i$, this is essentially a heat equation! In fact, for the real Schrödinger equation, the integral of $|u|^2$ is constant with time; any of the changes suggested would destroy this property, and then solutions could not even represent probability distributions. The fundamental equation of chemistry appears to itself be fine-tuned.

world that are vital to the existence of life and which contribute to the general impression of the improbability of the observed world."

Recently, some physicists have proposed a modification to the standard big bang model, which postulates a very short period of very rapid expansion of the universe near the beginning of time, and proposes to explain the fine tuning in the uniformity of the material density of the universe and another initial condition fine-tuning issue, called the "flatness problem." The details of this "inflationary universe" theory, which apparently requires some fine tuning itself, are still very speculative. One of the architects of the theory, Alan Guth [Guth 1998], in discussing the hypothetical "inflaton" field, invented to drive the inflation, says, "It must be admitted, however, that the ad hoc addition of such a field makes the theory look a bit contrived. To be honest, a theory of this sort *is* contrived, with the goal of arranging for the density perturbations to come out right." According to the authors of *Current Issues in Cosmology* [Pecker and Narlikar 2006], "Inflation, on the other hand, is still speculative.... What is dramatically missing here is the identification of the inflaton field."

Although Davies recognizes that some may see design in the fortuitous features of our universe, he attempts to defend the multiple universes theory. "If we believe that there are countless other universes, either in space or time, or in superspace, there is no longer anything astounding about the enormous degree of cosmic organization that we observe. We have selected it by our very existence. The world is just an accident that was bound to happen sooner or later," he says. Davies compares the anthropic principle's explanation of why the laws, particles and forces of physics are so friendly toward life to the traditional scientific explanation of why conditions on Earth are so ideally suited for life: "The many universes theory does provide an explanation for why many things around us are the way they are. Just as we can explain why we are living on a planet near a stable star by pointing out that only in such locations can life form, so we can perhaps explain many of the more general features of the universe by this anthropic selection process."

As Michael Behe points out in *The Edge of Evolution* [Behe 2007], however, anthropic selection only claims to explain why we live in a universe which can support life, it does not explain why we live in such a "lush" universe, where the fundamental laws of physics not only make life possible, they make it interesting. For example, some of the heavier chemical elements (for example, copper or uranium), which are probably not vital for life itself, have played a critical role in the progress of science and technology, and the existence and useful chemical properties of these elements can also be traced to the fine tuning of our physical laws.

A related argument is now being made regarding the "privileged" position of our planet. It is well-documented that the conditions on Earth are very fine-tuned for the development of life: our planet is the right size, with the right kind of atmosphere, it circles the right kind of star, it is the right distance from this star, to name only the most obvious. Of course, there are many stars, so it has always been argued that there were bound to be some planets in this huge universe with the conditions needed for life. (In fact, there is now some doubt about this, as research continually increases the known list of privileges enjoyed by our planet [Ward and Brownlee 2003].) But now some scientists, such as astronomer Guillermo Gonzalez [Gonzalez and Richards 2004], argue that our planet enjoys other "privileges" which are rare in the universe, which have nothing to do with survival, but seem to give us an ideal platform from which to *view* the universe.

According to the picture drawn by the popular media, primitive man attributed many phenomena in Nature to design, but science has progressively removed the need for the design hypothesis from these phenomena one by one, and now a group of religious fanatics is trying to make a last stand in biological origins, where things are most difficult to explain. The true picture is very different; in fact, we are discovering that primitive man was *not* wrong in attributing many natural phenomena to design, the design just dates back much farther than he imagined, to the origin of the universe. And of course all of the arguments in this chapter take for granted that once the right conditions to support life are present, life can spontaneously develop, an assumption for which there is absolutely no supporting evidence. As Richard Dawkins famously admitted in the movie *Expelled*,[3] no one really has any idea how life could have originated.

It is difficult to argue with those who appeal to "anthropic selection" to explain improbable circumstances; about all you can say is that there is a simpler explanation. But other universes are by definition beyond observation, so that the anthropic principle is untestable, and therefore unscientific. It is interesting to see how those who for many years have criticized the creationists for inventing an agent external to our universe to account for the appearance of man are now reduced to inventing other universes to explain our existence.

---

3. www.expelledthemovie.com

Fred Heeren [Heeren 1995] illustrates the silliness of the idea that, given enough universes, everything will eventually happen. If there are enough universes, he says, one of them would be just like ours except that in that one Elvis Presley kicked his drug habit, got involved in Tennessee politics, and became president of the United States. It seems much simpler to believe that there is only one universe, and it appears to be cleverly designed because it is cleverly designed.

2.2 Design in Mathematics

When we think of design, we normally think of biology, or perhaps physics, but usually not mathematics. How can we see design in something that could not be any different than it is? I don't know if mathematics could have been different than it is, but as a mathematician, I still see design in mathematics, and plenty of it. The non-mathematician, who may think of mathematics as consisting only of arithmetic and perhaps algebra and geometry, could never imagine the richness that is there, waiting to be discovered, in the many fields and subfields of mathematics. How could he imagine that there are enough interesting and challenging problems to keep thousands of mathematicians busy and entertained throughout their lifetimes? Number theory is the study of the positive integers: 1,2,3,4,.... One might think that at least in this field of mathematics the number of interesting problems would be soon exhausted, and that all of the most important problems would be quickly solved; but one would be quite wrong on both counts. Many simple-sounding problems in number theory remain unsolved to this day, for example, are there an infinite number of "twin primes"—primes which differ by 2 (the minimum possible), such as 881 and 883?

To illustrate the richness that can be hidden in a simple-looking mathematical definition or equation, I offer the following example, which does not require any advanced mathematics to appreciate. Consider the iteration:

$$x_{n+1} = ax_n(1 - x_n)$$

If a is between 1 and 3, you can start anywhere in the interval $0 < x_0 < 1$ and you will converge eventually to a specific fixed point, $x = \frac{a-1}{a}$. If a is between 3 and 3.4495, you can start almost anywhere (anywhere except exactly at the fixed point $x = \frac{a-1}{a}$, which no longer attracts nearby iterates) in the interval $(0, 1)$ and you will end up oscillating back and forth between two specific points, that is, you will converge to an "orbit" of period 2. If you increase a a little more, you can start almost anywhere (anywhere except at the fixed point or one of the points of the period 2 orbit, which is also no longer "attractive") and you will converge to an orbit of period 4, a little more and the attractive orbits have higher and higher periods. Finally, if a is increased to 4, you have the following strange situation. There are now orbits of every possible period, but none of them are attractive, so if your x_0 does not belong to any of these orbits, you will wander around in a random-looking, "chaotic," fashion forever, never converging to any fixed point or any periodic orbit. Since there are an infinite number of periodic orbits, one might think that if you pick x_0 at random your chances of missing all of them would be pretty low. But in fact, your chances are 100%—you are virtually certain to miss them all, and jump around in a chaotic manner forever. All of this information can be gleaned from analyzing this simple equation, called the logistic equation.

Biologists find that as they look deeper and deeper into the cell, things do not become simpler, they become more complex, and more interesting. Scientists in other fields have also found that the deeper they dig, the more interesting things become, and I see no sign that any subfield of mathematics is beginning to run dry either, all indications are that there are as many entertaining problems in each of them as there are mathematicians to study them. Furthermore, the connection between mathematics and the sciences is truly astounding, and points to a common designer. Who could have imagined, for example, that a field of "pure" mathematics such as algebraic group theory would be important to quantum physics? Similar mathematical equations show up in the most diverse applications. To cite an example in my own area of expertise: partial differential equations find applications in almost every conceivable field of science and engineering, as can be appreciated by looking at the list of publications (in over 100 *different* journals, at www.vni.com/pde2d) in which my partial differential equation-solving software has been used to simulate physical phenomena.

British physicist Sir James Jeans said "From the intrinsic evidence of his creation, the Great Architect of the Universe begins to appear as a pure mathematician." I don't know how mathematics could have been different than it is, but I nevertheless insist: mathematics is also designed.

2.3 Supplement: The Stability of Planetary Orbits

In our universe, the force of gravity between two bodies of masses M and m, at a distance r apart, is given by $F = -GMm/r^2$, where G is the gravitational constant. We have already mentioned that if we play with the constant G and make it a bit larger or smaller, terrible things would happen that make it impossible for our universe to support life, but it requires some advanced physics to show this. However, it requires only a little physics, and a little calculus, to see what would happen if we play with the r^2 term in the denominator, so let's do this. The results are not nearly as striking, but this is one of the few examples of "fine-tuning" that can be understood without any advanced physics.

So let us replace our inverse square law of gravity by $F = -GMm/r^n$, where n may be something other than 2, and let us look at the orbit of the Earth around the sun. If the position of our sun, of mass M, is taken to be fixed at the origin, and the position of Earth, of mass m, is given by $(x(t), y(t), z(t))$, Newton's second law says

$$m(x'', y'', z'') = -\frac{GMm}{r^n}(\frac{x}{r}, \frac{y}{r}, \frac{z}{r})$$

that is, the mass of the Earth times its acceleration vector is equal to the force of gravity on the Earth, which is a vector of magnitude GMm/r^n in the direction of the unit vector $-(\frac{x}{r}, \frac{y}{r}, \frac{z}{r})$, ie, toward the sun.

Since orbits will remain in the plane they start in, and we can take the z axis to be normal to this plane, we can express the Earth's position using polar coordinates as $x(t) = r\ cos(\theta), y(t) = r\ sin(\theta), z(t) = 0$. Now, after a bit of work, the above differential equations can be written in polar coordinates as:

$$\begin{aligned} r'' - r(\theta')^2 &= -GM/r^n \\ 2r'\theta' + r\theta'' &= 0 \end{aligned}$$

The second equation, after multiplying through by r, is equivalent to $(r^2\theta')' = 0$, which means $r^2\theta' = c$, a constant. Substituting $\theta' = c/r^2$ into the first equation, we get a differential equation for $r(t)$:

$$r'' = c^2/r^3 - GM/r^n \tag{2.1}$$

From equation (2.1) we can see why orbits can be stable in an inverse-square force field: if $n = 2$, then when r gets too small, the positive term (due to the centrifugal force) dominates, and the radial acceleration is positive, which tends to increase r. When r gets too large, the negative term (due to gravity) dominates, and the radial acceleration is back toward the sun. But what if we increase n, to 3? Now, $r'' = (c^2 - GM)/r^3$ and if $c^2 - GM$ is positive the acceleration will always be positive, and the Earth would spiral away out of the solar system; if $c^2 - GM$ is negative, the Earth would spiral into the sun. Neither outcome would be very healthy for life on Earth! If n is even larger than 3, the negative term in (2.1) dominates when r is small, and the positive term dominates when r is large, so that all orbits of all planets are again unstable.

Orbits can still be stable if we decrease n, to 1. But now the potential energy for an object of mass m, associated with the Earth's gravitational field, would be $V(r) = GMm\ ln(r)$, where M is the mass of the Earth. Note that the potential energy at $r = \infty$ would be infinite (in an inverse square field it is finite). This means there would be no theoretical limit to the energy with which a meteor or other object of a given mass could hit the Earth. Exploration of deep space would also be difficult, obviously.

Now one could argue that it is only natural that in our three-dimensional universe, gravity would obey an inverse-square law. In an N-dimensional universe, the energy from a source (e.g., the sun) is spread out, at a distance r, over an "area" of size proportional to $1/r^{N-1}$, so its intensity is proportional to $1/r^{N-1}$. Thus it might seem reasonable that the effect of the sun's gravity would also die out at this rate as we move away from it. But if we accept this argument, we can say we are lucky we live in a three-dimensional universe, because if N were 4 or more, gravity would obey an inverse cube (or worse) law, and since orbits would still be planar (a planet would remain in the 2D subspace spanned by its initial position and velocity vectors), the above polar coordinate analysis is still valid and shows that all orbits would be unstable in universes of more than three dimensions. And who wants to live in a 1D or 2D universe, where all we could see would be points or lines!

3

A Mathematician's View of Evolution

The following article appeared in ***The Mathematical Intelligencer*** [Sewell 2000].

Used with kind permission of Springer Science and Business Media.

3.1 Darwin's Black Box

In 1996, Lehigh University biochemist Michael Behe published a book entitled *Darwin's Black Box* [Behe 1996], whose central theme is that every living cell is loaded with features and biochemical processes which are "irreducibly complex"—that is, they require the existence of numerous complex components, each essential for function. Thus, these features and processes cannot be explained by gradual Darwinian improvements, because until all the components are in place, these assemblages are completely useless, and thus provide no selective advantage. Behe spends over 100 pages describing some of these irreducibly complex biochemical systems in detail, then summarizes the results of an exhaustive search of the biochemical literature for Darwinian explanations. He concludes that while biochemistry texts often pay lip-service to the idea that natural selection of random mutations can explain everything in the cell, such claims are pure "bluster," because "there is no publication in the scientific literature that describes how molecular evolution of any real, complex, biochemical system either did occur or even might have occurred."

When Dr. Behe was at the University of Texas El Paso in May of 1997 to give an invited talk, I told him that I thought he would find more support for his ideas in mathematics, physics and computer science departments than in his own field. I know a good many mathematicians, physicists and computer scientists who, like me, are appalled that Darwin's explanation for the development of life is so widely accepted in the life sciences. Few of them ever speak out or write on this issue, however—perhaps because they feel the question is simply out of their domain. However, I believe there are two central arguments against Darwinism, and both seem to be most readily appreciated by those in the more mathematical sciences.

3.2 Irreducible Complexity

The cornerstone of Darwinism is the idea that major (complex) improvements can be built up through many minor improvements; that the new organs and new systems of organs which gave rise to new orders, classes and phyla developed gradually, through many very minor improvements. We should first note that the fossil record does not support this idea, for example, Harvard paleontologist George Gaylord Simpson [Simpson 1960] writes:

> It is a feature of the known fossil record that most taxa appear abruptly. They are not, as a rule, led up to by a sequence of almost imperceptibly changing forerunners such as Darwin believed should be usual in evolution.... This phenomenon becomes more universal and more intense as the hierarchy of categories is ascended. Gaps among known species are sporadic and often small. Gaps among known orders, classes and phyla are systematic and almost always large. These peculiarities of the record pose one of the most important theoretical problems in the whole history of life: Is the sudden appearance of higher categories a phenomenon of evolution or of the record only, due to sampling bias and other inadequacies?

An April, 1982, *Life* magazine article (excerpted from Francis Hitching's book, *The Neck of the Giraffe: Where Darwin Went Wrong*) contains the following report:

> When you look for links between major groups of animals, they simply aren't there.... 'Instead of finding the gradual unfolding of life,' writes David M. Raup, a curator of Chicago's Field Museum of Natural History, 'what geologists of Darwin's time and geologists of the present day actually find is a highly uneven or jerky record; that is, species appear in the fossil sequence very suddenly, show little or no change during their existence, then abruptly disappear.' These are not negligible gaps. They are peri-

> ods, in all the major evolutionary transitions, when immense physiological changes had to take place.

Even among biologists, the idea that new organs, and thus higher categories, could develop gradually through tiny improvements has often been challenged. How could the "survival of the fittest" guide the development of new organs through their initial useless stages, during which they obviously present no selective advantage? (This is often referred to as the "problem of novelties.") Or guide the development of entire new systems, such as nervous, circulatory, digestive, respiratory and reproductive systems, which would require the simultaneous development of several new interdependent organs, none of which is useful, or provides any selective advantage, by itself? French biologist Jean Rostand, for example, wrote [Rostand 1956]:

> It does not seem strictly impossible that mutations should have introduced into the animal kingdom the differences which exist between one species and the next... hence it is very tempting to lay also at their door the differences between classes, families and orders, and, in short, the whole of evolution. But it is obvious that such an extrapolation involves the gratuitous attribution to the mutations of the past of a magnitude and power of innovation much greater than is shown by those of today.

Behe's book is primarily a challenge to this cornerstone of Darwinism at the microscopic level. Although we may not be familiar with the complex biochemical systems discussed in this book, I believe mathematicians are well qualified to appreciate the general ideas involved.

And although an analogy is only an analogy, perhaps the best way to understand Behe's argument is by comparing the development of the genetic code of life with the development of a computer program. Suppose an engineer attempts to design a structural analysis computer program, writing it in a machine language that is totally unknown to him. He simply types out random characters at his keyboard, and periodically runs tests on the program to recognize and select out chance improvements when they occur. The improvements are permanently incorporated into the program while the other changes are discarded. If our engineer continues this process of random changes and testing for a long enough time, could he eventually develop a sophisticated structural analysis program? (Of course, when intelligent humans decide what constitutes an "improvement," this is really artificial selection, so the analogy is far too generous.)

If a billion engineers were to type at the rate of one random character per second, there is virtually no chance that any one of them would, given the 4.5 billion year age of the Earth to work on it, accidentally duplicate a given 20-character improvement. Thus our engineer cannot count on making any major improvements through chance alone. But could he not perhaps make progress through the accumulation of very small improvements? The Darwinist would presumably say, yes, but to anyone who has had minimal programming experience this idea is equally implausible. Major improvements to a computer program often require the addition or modification of hundreds of interdependent lines, no one of which makes any sense, or results in any improvement, when added by itself. Even the smallest improvements usually require adding several new lines. It is conceivable that a programmer unable to look ahead more than 5 or 6 characters at a time might be able to make some very slight improvements to a computer program, but it is inconceivable that he could design anything sophisticated without the ability to plan far ahead and to guide his changes toward that plan.

If archeologists of some future society were to unearth the many versions of my PDE solver, PDE2D, which I have produced over the last 20 years, they would certainly note a steady increase in complexity over time, and they would see many obvious similarities between each new version and the previous one. In the beginning it was only able to solve a single linear, steady-state, 2D equation in a polygonal region. Since then, PDE2D has developed many new abilities: it now solves nonlinear problems, time-dependent and eigenvalue problems, systems of simultaneous equations, and it now handles general curved 2D regions. Over the years, many new types of graphical output capabilities have evolved, and in 1991 it developed an interactive preprocessor, and more recently PDE2D has adapted to 3D and 1D problems. An archeologist attempting to explain the evolution of this computer program in terms of many tiny improvements might be puzzled to find that each of these major advances (new classes or phyla??) appeared suddenly in new versions; for example, the ability to solve 3D problems first appeared in version 4.0. Less major improvements (new families or orders??) appeared suddenly in new subversions, for example, the ability to solve 3D problems with periodic boundary conditions first appeared in version 5.6. In fact, the record of PDE2D's development would be similar to the fossil record, with large gaps where major new features appeared, and smaller gaps where minor ones appeared. That is because the multitude of intermediate programs between versions or subversions which the archeologist might expect to find never existed, because—for example—none of the changes I made for edition 4.0 made any sense, or provided PDE2D any advantage whatever in solving 3D problems (or anything else) until hundreds of lines had been added.

Whether at the microscopic or macroscopic level, major, complex, evolutionary advances, involving new features (as opposed to minor, quantitative changes such as an increase in the length of the giraffe's neck,[1] or the darkening of the wings of a moth, which clearly could occur gradually) also involve the addition of many interrelated and interdependent pieces. These complex advances, like those made to computer programs, are not always "irreducibly complex"— sometimes there are intermediate useful stages. But just as major improvements to a computer program cannot be made 5 or 6 characters at a time, certainly no major evolutionary advance is reducible to a chain of tiny improvements, each small enough to be bridged by a single random mutation.

3.3 The Second Law of Thermodynamics

The other point is very simple, but also seems to be appreciated only by more mathematically-oriented people. It is that to attribute the development of life on Earth to natural selection is to assign to it—and to it alone, of all known natural "forces"—the ability to violate the sec-

1. Ironically, W. E. Lönnig's article "The Evolution of the Long-Necked Giraffe," (www.weloennig.de/Giraffe.pdf) has since convinced me that even this is far beyond the ability of natural selection to explain.

ond law of thermodynamics and to cause order to arise from disorder. It is often argued that since the Earth is not a closed system—it receives energy from the Sun, for example—the second law is not applicable in this case. It is true that order can increase locally, if the local increase is compensated by a decrease elsewhere, i.e., an open system can be taken to a less probable state by importing order from outside. For example, we could transport a truckload of encyclopedias and computers to the moon, thereby increasing the order on the moon, without violating the second law. But the second law of thermodynamics—at least the underlying principle behind this law—simply says that natural forces do not cause extremely improbable things to happen,[2] and it is absurd to argue that because the Earth receives energy from the Sun, this principle was not violated here when the original rearrangement of atoms into encyclopedias and computers occurred.

The biologist studies the details of natural history, and when he looks at the similarities between two species of butterflies, he is understandably reluctant to attribute the small differences to the supernatural. But the mathematician or physicist is likely to take the broader view. I imagine visiting the Earth when it was young and returning now to find highways with automobiles on them, airports with jet airplanes, and tall buildings full of complicated equipment, such as televisions, telephones and computers. Then I imagine the construction of a gigantic computer model which starts with the initial conditions on Earth 4 billion years ago and tries to simulate the

2. An unfortunate choice of words; I should have said, the underlying principle behind the second law is that natural forces do not do *macroscopically* describable things which are extremely improbable from the *microscopic* point of view.

effects that the four known forces of physics (the gravitational, electromagnetic and strong and weak nuclear forces[3]) would have on every atom and every subatomic particle on our planet (perhaps using random number generators to model quantum uncertainties!). If we ran such a simulation out to the present day, would it predict that the basic forces of Nature would reorganize the basic particles of Nature into libraries full of encyclopedias, science texts and novels, nuclear power plants, aircraft carriers with supersonic jets parked on deck, and computers connected to laser printers, CRTs and keyboards? If we graphically displayed the positions of the atoms at the end of the simulation, would we find that cars and trucks had formed, or that supercomputers had arisen? Certainly we would not, and I do not believe that adding sunlight to the model would help much. Clearly something extremely improbable has happened here on our planet, with the origin and development of life, and especially with the development of human consciousness and creativity.

3. "One of the most remarkable simplifications in physics is that only four distinct forces account for all known phenomena,"—*College Physics* [Urone 2001].

4

Postscript in 1985 Book

The following appeared as a "Postscript" in my 1985 book, ***Analysis a Finite Element Method: PDE/PROTRAN*** *[Sewell 1985]. It is very similar to the first part of the* ***Mathematical Intelligencer*** *article reproduced in the previous chapter, but is of some historical interest because of the early date.*

As I begin my 12th year of work on TWODEPEP (now PDE/PROTRAN), I am intrigued by the analogy between the 11-year evolution of this computer code and the multi-billion year history of the genetic code of life, which contains a blueprint for a species encoded into billions of bits of information. Like the code of life, TWODEPEP began with primitive features, being capable of solving only a single linear elliptic equation in polygonal regions, with simple boundary conditions. It passed through many useful stages as it adapted to non-linear and time dependent problems, systems of PDEs, eigenvalue problems, and as it evolved cubic and quartic elements and isoparametric elements for curved boundaries. It grew a preprocessor and a graphical output package, and out-of-core frontal and conjugate gradient methods were added to solve the linear systems.

Each of these changes represented major evolutionary steps—new orders, classes or phyla, if you will. The conjugate gradient method, in turn, also passed through several less major variations as the basic method was modified to precondition the matrix, to handle nonsymmetric systems, and as stopping criteria were altered, etc. Some of these variations might be considered new families, some new genera, and some only special changes.

I see one flaw in the analogy, however. While I am told that the DNA code was designed by a natural process capable of recognizing improvements but incapable of planning beyond the next random mutation, I find it difficult to believe that TWODEPEP could have been designed by a programmer incapable of thinking ahead more than a few characters at a time.

But perhaps, it might be suggested, a programmer capable of making only random changes, but quite skilled at recognizing improvements could, given 4.5 billion years to work on it, evolve such a program. A few simple calculations would convince him that this programmer would have to rely on very tiny improvements. For example, if he could produce a billion random "mutations" per second (or, for a better analogy, suppose a billion programmers could produce one "mutation" per second each), he could not, statistically, hope to produce any predetermined 20 character improvement during this time period. Could such a programmer, with no programming or mathematical skills other than the ability to recognize and select out very small improvements through testing, design a sophisticated finite element program?

The Darwinist would presumably say, yes, but to anyone who has had minimal programming experience such an idea is preposterous. The major changes to TWODEPEP, such as the addition of a new linear equation solver or new element, required the addition or modification of hundreds of lines of code before the new feature was functional. None of the changes made during this period were of any use whatever until all were in place.

Even the smallest modifications to that new feature, once it was functional, required adding several lines, no one of which made any sense, or provided any "selective advantage," when added by itself.

Consider, by way of analogy, the water-tight trap of the carnivorous bladderwort plant, which has a double sealed, valve-like door which is opened when a trigger hair is activated, causing the victim to be sucked into the vacuum of the trap (described by R.F.Daubenmire in *Plants and Environment,* [Daubenmire 1947]). It is difficult to see what selective advantage this trap provided until it was almost perfect.

This, then, is the fallacy of Darwin's explanation for the causes of evolution—the idea that major (complex) improvements can be broken down into many minor improvements. French biologist Jean Rostand, in *A Biologist's View* [Rostand 1956] recognized this:

> It does not seem strictly impossible that mutations should have introduced into the animal kingdom the differences which exist between one species and the next... hence it is very tempting to lay also at their door the differences between classes, families and orders, and,

> in short, the whole of evolution. But it is obvious that such an extrapolation involves the gratuitous attribution to the mutations of the past of a magnitude and power of innovation much greater than is shown by those of today.

The famous "problem of novelties" is another formulation of the objection raised here. How can natural selection cause new organs to arise and guide their development through the initial stages during which they present no selective advantage, the argument goes. The Darwinist is forced to argue that there are no useless stages. He believes that new organs and new systems of organs arose gradually, through many small improvements. But this is like saying that TWODEPEP could have made the transition from a single PDE to systems of PDEs through many five or six character improvements, each of which made it work slightly better on systems.

It is interesting to note that this belief is not supported even by the fossil evidence. Harvard paleontologist George Gaylord Simpson, for example, in "The History of Life," Volume II of *Evolution after Darwin,* [Simpson 1960] points out:

> It is a feature of the known fossil record that most taxa appear abruptly. They are not, as a rule, led up to by a sequence of almost imperceptibly changing forerunners such as Darwin believed should be usual in evolution.... This phenomenon becomes more universal and more intense as the hierarchy of categories is ascended. Gaps among known species are sporadic and often small. Gaps among known orders, classes and phyla are systematic and almost always large. These peculiarities of the

> record pose one of the most important theoretical problems in the whole history of life: Is the sudden appearance of higher categories a phenomenon of evolution or of the record only, due to sampling bias and other inadequacies?

Another way of describing this same structure is expressed in a recent *Life* magazine article (Francis Hitching, "Was Darwin Wrong on Evolution?", April 1982, which concludes that "natural selection has been tested and found wanting") which focuses on the "curious consistency" of the fossil gaps:

> These are not negligible gaps. They are periods, in all the major evolutionary transitions, when immense physiological changes had to take place.

Unless we are willing to believe that useless, "developing" organs (and insect traps which could almost catch insects) abounded in the past, we should have expected the fossil structure outlined above, with large gaps between the higher categories, where new organs and new systems of organs appeared.

Nevertheless, despite the fact that the structure of the fossil record is the only argument against Darwin which has received much attention lately, this is not the real issue. The "problem of novelties" correctly states the real argument, but too weakly. Consider, for example, the human eye, with an aperture whose size varies automatically according to the light intensity, controlled by reflex signals from the brain; with a lens whose curvature varies automatically according to the distance to the object in view; and with a retina which receives the picture on color sensitive cells and transmits it, complete with coded intensity and frequency information, through the optic nerve to the brain. The brain superimposes the pictures from the two eyes and stores this 3D picture somehow in memory, and it will be able to search for and recall this image later and use it to recognize an older but familiar face in a different picture. Like TWODEPEP, the eye has passed through various useful stages in its development, but it contains a large number of features which could not reach usefulness in a single random mutation and which provided no selective advantage until useful (e.g. the nerves and arteries which service it), and many groups of features which are useless individually. The Darwinist may bridge the gaps between taxa with a long chain of tiny improvements in his imagination, but the analogy with software puts his ideas into perspective. The idea that all the magnificent species in the living world, or the human brain with its human consciousness, could have arisen from simple organic molecules guided by a natural process unable to plan beyond the next tiny mutation, is entirely comparable to the idea that a programmer incapable of thinking ahead more than a few characters at a time could, given a lot of time, design any sophisticated computer program.

I suggest that, with Jean Rostand, “we must have the courage to recognize that we know nothing of the mechanism” of evolution.

5

Can 'Anything' Happen in an Open System?

5.1 A Second Look at the Second Law

The following is a substantially-modified version of an article which appeared in the on-line ***American Spectator*** *(www.spectator.org/dsp_article.asp?art_id=9128), December 28, 2005.*

In the current debate over "intelligent design," the strongest argument offered by opponents of design is this: we have scientific explanations for most everything else in Nature, what is special about evolution? The layman understands quite well that explaining the appearance of human brains is a fundamentally different and much more difficult sort of problem from finding the causes of earthquakes; however, to express this difference in terms a scientist can understand requires a discussion of the second law of thermodynamics.

The first formulations of the second law were all about heat: a quantity called thermal "entropy" was defined to measure the randomness, or disorder, associated with a temperature distribution, and it was shown that in an isolated system this entropy always increases, or at least never decreases, as heat diffuses and the temperature becomes more and more randomly (more uniformly) distributed. If we define thermal "order" to be the opposite (negative) of thermal entropy, we can say that the thermal order can never increase in a closed (isolated) system. However, it was soon realized that other types of order can be defined which also never increase in a closed system, for example, we can define a "carbon order" associated with the distribution of carbon diffusing in a solid, using the same equations, and through an identical analysis show that this order also continually decreases, in a closed system. With time, the second law came to be interpreted more and more generally, and today most discussions of the second law in physics textbooks offer examples of entropy increases (or order decreases, since we are defining order to be the opposite of entropy[1]) which have nothing to do with heat conduction or diffusion, such as the shattering of a wine glass or the demolition of a building.

For example, in *Basic Physics* [Ford 1968] Kenneth Ford writes,

> Imagine a motion picture of any scene of ordinary life run backward. You might watch... a pair of mangled automobiles undergoing instantaneous repair as they back apart. Or a

1. "Entropy" sounds so much more scientific than "order," but if a measure is scientific, so is its negative!

> dead rabbit rising to scamper backward into the woods as a crushed bullet re-forms and flies backward into a rifle.... Or something as simple as a cup of coffee on a table gradually becoming warmer as it draws heat from its cooler surroundings. All of these backward-in-time views and a myriad more that you can quickly think of are ludicrous and impossible for one reason only—they violate the second law of thermodynamics. In the actual scene of events, entropy is increasing. In the time reversed view, entropy is decreasing.

It is a well-known prediction of the second law that, in a closed system, every type of order is unstable and must eventually decrease, as everything tends toward more probable states. Natural forces, such as corrosion, erosion, fire and explosions, do not create order, they destroy it. S. Angrist and L. Hepler, in *Order and Chaos* [Angrist and Hepler 1967], write, "An arsonist working on a big library is merely speeding up the inevitable result demanded by the second law."

The second law is all about probability, it uses probability at the microscopic level to predict macroscopic change: the reason carbon distributes itself more and more uniformly in an insulated solid is, that is what the laws of probability predict when diffusion alone is operative. The reason natural forces may turn a spaceship, or a TV set, or a computer into a pile of rubble but not vice-versa is also probability: of all the possible arrangements atoms could take, only a very small percentage could fly to the moon and back, or receive pictures and sound from the other side of the Earth, or add, subtract, multiply and divide real numbers with high accuracy. The second law is the reason that automobiles will degenerate into scrap metal over time (or quickly, as in Ford's movie) and, in the absence of intelligence, the reverse process will not occur; and it is the reason that Ford's rabbit and other animals, when they die, decay into simple organic and inorganic compounds, and, in the absence of intelligence, the reverse process will not occur.

The discovery that life on Earth developed through evolutionary "steps," coupled with the observation that mutations and natural selection—like other natural forces—can cause (minor) change, is widely accepted in the scientific world as proof that natural selection—alone among all natural forces—can create order out of disorder, and even design human brains, with human consciousness. Only the layman seems to see the problem with this logic. And where is the overwhelming evidence that justifies not only believing that natural selection can design human brains, but justifies branding as "anti-science" anyone who doubts that it can? In his new book, *The Edge of Evolution* [Behe 2007], Lehigh University biochemist Michael Behe looks in considerable detail at the struggle for survival between humans and the malaria parasite where, in the last 100 years, the evolution of far more organisms can be studied than were involved in the entire natural history of mammals. He finds that natural selection can be credited with some very minor change, but "Far and away the most extensive relevant data we have on the subject of evolution's effects on competing organisms is that accumulated on interactions between humans and our parasites. As with the example of malaria, the data show trench warfare, with acts of desperate destruction, not arms races, with mutual improvements. The thrust and parry of human-malaria evolution did not build anything—it only destroyed things." Behe also looks at Richard Lenski's 20-year *E.coli* experiment, which a June 9, 2008 *New Scientist* article now claims represents "the first time evolution has been caught in the act," and concludes that "nothing fundamentally new has been produced." Behe claims that the minor changes observed in this experiment are all due to "breaking some genes and turning others off." In any case, the *New Scientist* article contains a remarkable admission, that natural selection has never before (and not even now according to Behe) been actually observed to produce any significant advance! To claim that the mechanism which produces such minor changes in bacteria and parasite populations is capable of producing human brains is an incredible extrapolation, yet this claim is routinely presented as being as well-established as gravity. In *any* other field, a scientist making such an extrapolation with such confidence would be the laughingstock of his peers.

In a 2000 *Mathematical Intelligencer* article,[2] I asserted that the idea that the four fundamental forces of physics alone could rearrange the fundamental particles of Nature into spaceships, nuclear power plants, and computers, connected to laser printers, CRTs, keyboards and the internet, appears to violate the second law of thermodynamics in a spectacular way.

Anyone who has made such an argument is familiar with the standard reply: the Earth is an open system, it receives energy from the sun, and order can increase in an open system, as long as it is "compensated" somehow by a comparable or greater decrease outside the system. Peter Urone, for example, in *College Physics* [Urone 2001] writes, "Some people misuse the second law of thermo-

2. Chapter 3

dynamics, stated in terms of entropy, to say that the existence and evolution of life violate the law and thus require divine intervention.... It is true that the evolution of life from inert matter to its present forms represents a large decrease in entropy for living systems. But it is *always* possible for the entropy of one part of the universe to decrease, provided the total change in entropy of the universe increases."

According to this reasoning, then, the second law does not prevent scrap metal from reorganizing itself into a computer in one room, as long as two computers in the next room are rusting into scrap metal—and the door is open. (Or the thermal order in the next room is decreasing—though I'm not sure what the conversion rate is between computers and thermal order!) This strange argument of "compensation" makes no sense logically: an extremely improbable event is not rendered less improbable by the occurrence of other events which are more probable. To understand where this argument of compensation comes from, one needs to understand that of the example applications mentioned in the Ford text above, the coffee cup example is special: the application to heat conduction is special not only because it was the first application, but because it is quantifiable. It is commonly used as the "model" problem on which our thinking about the other, less quantifiable, applications is based. The fact that thermal order cannot increase in a closed system, but can increase in an open system, was used to conclude that, in other applications, anything can happen in an open system as long as it is compensated by order decreases outside this system, so that the total "order" in the universe (or any closed system containing the open system) still decreases.

In Appendix D of *The Numerical Solution of Ordinary and Partial Differential Equations* [Sewell 2005], included as section 5.6, I take a closer look at the equations for entropy change, which apply not only to thermal entropy but also to the entropy associated with anything else that diffuses, and show that they do not simply say that order cannot increase in a closed system, they also say that in an open system, order cannot increase faster than it is imported through the boundary. According to these equations, the thermal order in an open system can decrease in two different ways—it can be converted to disorder, or it can be exported through the boundary. It can increase in only one way: by importation through the boundary. Similarly, the increase in "carbon order" in an open system cannot be greater than the carbon order imported through the boundary, and the increase in "chromium order" cannot be greater than the chromium order imported through the boundary, and so on.

The "compensation" argument was produced by people who generalized the model equation for closed systems, but forgot to generalize the equation for open systems. Both equations are only valid for our simple models, where it is assumed that only heat conduction or diffusion is going on; naturally in more complex situations, the laws of probability do not make such simple predictions. Nevertheless, in [Sewell 2001] I generalized the equation for open systems to the following tautology, which is valid in all situations: *"If an increase in order is extremely improbable when a system is closed, it is still extremely improbable when the system is open, unless something is entering which makes it* not *extremely improbable."* The fact that order is disappearing in the next room does not make it any easier for computers to appear in our room—unless this order is disappearing *into* our room, and then only if it is a type of order that makes the appearance of computers not extremely improbable, for example, computers. Importing thermal order will make the temperature distribution less random, and importing carbon order will make the carbon distribution less random, but neither makes the formation of computers more probable.

What happens in a closed system depends on the initial conditions; what happens in an open system depends on the boundary conditions as well. As I wrote in "Can ANYTHING Happen in an Open System?" [Sewell 2001] "order can increase in an open system, not because the laws of probability are suspended when the door is open, but simply because order may walk in through the door.... If we found evidence that DNA, auto parts, computer chips, and books entered through the Earth's atmosphere at some time in the past, then perhaps the appearance of humans, cars, computers, and encyclopedias on a previously barren planet could be explained without postulating a violation of the second law here.... But if all we see entering is radiation and meteorite fragments, it seems clear that what is entering through the boundary cannot explain the increase in order observed here."

5.2 Many Types of Order

Evolution is a movie running backward, that is what makes it so different from every other known process in our universe, and that is why it demands a radically different explanation. Evolutionists have always countered this argument with the claim that the transformation of a barren, rocky planet into what we see today does not violate the second law because the Earth receives energy from the sun. Now that the silliness of this argument has become evident we are beginning to hear a new argument, which basically says, wait a minute, the second law of thermodynamics really applies only to thermodynamics after all, because otherwise, there are so many types of order that order is a meaningless concept.

In the original *Mathematical Intelligencer* article [Sewell 2000] I made the assertion that the underlying principle behind the second law is that natural forces do not do extremely improbable things. The journal and I received several replies arguing that everything Nature does can be considered extremely improbable—the exact arrangement of atoms at any time at any place is extremely unlikely to be repeated, noted one e-mail. In another published reply [Davis 2001], the author made an analogy with coin flipping and argued that any particular sequence of heads and tails is extremely improbable, so something extremely improbable happens every time we flip a long series of coins. If a coin were flipped 1000 times, he would apparently be no more surprised by a string of all heads than by any other sequence, because any string is as improbable as another. This critic concedes that it is extremely unlikely that humans and computers would arise again if history were repeated, "but something would."

Obviously, I should have been more careful with my wording in the first article: I should have said that the underlying principle behind the second law is that natural forces do not do *macroscopically* describable things which are extremely improbable from the *microscopic* point of view. A "macroscopically describable" event is just any event which can be described without resorting to an atom-by-atom (or coin-by-coin) accounting. Carbon distributes itself more and more uniformly in an insulated solid because there are many more arrangements of carbon atoms which produce nearly uniform distributions than produce highly nonuniform distributions. Natural forces may turn a spaceship into a pile of rubble, but not vice-versa—not because the exact arrangement of atoms in a given spaceship is more improbable than the exact arrangement of atoms in a given pile of rubble, but because (whether the Earth receives energy from the Sun or not) there are very few arrangements of atoms which would be able to fly to the moon and return safely, and very many which could not. The reader familiar with William Dembski's "specified complexity" concept [Dembski 2006], will recognize similarities to the argument here: natural forces do not do things which are "specified" (macroscopically describable) and "complex" (extremely improbable). Both are just attempts to state in more "scientific" terms what is already obvious to the layman, that unintelligent forces cannot do intelligent things.

Another popular way to state this: only intelligence can create information. For example, Casey Luskin of the Discovery Institute, in a December 2, 2008 entry at www.evolutionnews.org, reports finding a children's book about life on other planets, which provides a "recipe" for life: organic molecules, water and energy. Luskin points out that "One extremely important component that is missing from our 'recipe' [is] information... experience teaches that the sort of information we find in life... has only one real common source: intelligent agency." Molecular biologist Jonathan Wells [Wells 2006] says "The secret of life is not the physical DNA molecule, but the information it carries." Stephen Meyer, in his new book, *Signature in the Cell* [Meyer 2009], says "information typically degrades over time unless intelligent agents generate (or regenerate) it." The second law of thermodynamics applies to information!

If we toss a billion (fair) coins, it is true that any sequence is as improbable as any other, but most of us would still be surprised, and suspect that something other than chance is going on, if the result were "all heads," or "alternating heads and tails," or even "all tails except for coins $3i + 5$, for integer i." When we produce simply describable results like these, we have done something "macroscopically" describable which is extremely improbable. There are so many simply describable sequences possible that it is tempting to think that all or most outcomes could be simply described in some way, but in fact, there are fewer than 2^{30000} different 1000-word paragraphs, so the odds are about $2^{999970000}$ to 1 that a given sequence will not be that highly ordered—so our surprise would be quite justified. And if it can't be described in 1000 English words and symbols, it isn't very simply describable.

In the real world it is sometimes much harder to say what the laws of probability predict than in a coin-flipping experiment; thus here it may be even harder to define and measure order, but sometimes it is easy. In any case, with 10^{23} molecules in a mole of anything, we can be confident that the laws of probability at the microscopic level will be obeyed (at least on planets without life) as they apply to *all* macroscopic phenomena; this is precisely the assumption—the *only* common thread—behind all applications of the second law. Everything the second law predicts, it predicts with such high probability that it is as reliable as any other law of science—tossing a billion heads in a row is child's play compared to appreciably violating the second law in any application. One critic [Rosenhouse 2001] wrote "His claim that 'natural forces do not cause extremely improbable things to happen' is pure gibberish. Does Sewell invoke supernatural forces to explain the winning numbers in last night's lottery?" But getting the right number on 5 or 6 balls is not extremely improbable, in thermodynamics "extremely improbable" events involve getting the "right number" on 100,000,000,000,000,000,000,000 or so balls! If every atom on Earth bought one ticket every second since the big bang (about 10^{70} tickets) there is virtually no chance that any would ever win even a 100-ball lottery, much less this one. And since the second law derives its authority from logic alone, and thus cannot be overturned by future discoveries, Sir Arthur Eddington [Eddington 1929] called it the "supreme" law of Nature.

Although it is true that we sometimes are not sure *what* the second law predicts, it is not true that there are so many macroscopically describable phenomena that the second law cannot be expected to hold when applied to all of them—there are *relatively* few simply describable phenomena. It is not true, as the new argument asserts, that there are so many types of order that the appearance of computers and TV sets needs no explanation.

5.3 Darwin's Order Source

The evolutionist, therefore, cannot avoid the question of probability by saying that anything can happen in an open system, nor can he avoid it by saying that there are so many types of order that order is a meaningless concept. He is finally forced to argue that it only seems extremely improbable, but really isn't, that atoms would rearrange themselves into spaceships and computers and the internet.

Darwinists believe they have already discovered the source of all this order, so let us look more closely at their theory. The traditional argument against Darwinism is that natural selection cannot guide the development of new organs and new systems of organs—i.e., the development of new orders, classes and phyla—through their initial useless stages, during which they provide no selective advantage. Natural selection may be able to darken the wings of a moth (even this is disputed), but that does not mean it can design anything complex.

Consider, for example, the aquatic bladderwort, described in *Plants and Environment* [Daubenmire 1947]:

> The aquatic bladderworts are delicate herbs that bear bladder-like traps 5mm or less in diameter. These traps have trigger hairs attached to a valve-like door which normally keeps the trap tightly closed. The sides of the trap are compressed under tension, but when a small form of animal life touches one of the trigger hairs the valve opens, the bladder suddenly expands, and the animal is sucked into the trap. The door closes at once, and in about 20 minutes the trap is set ready for another victim.

In a *Nature Encyclopedia of Life Sciences* [Lönnig and Becker 2004] article on Carnivorous Plants, authors Wolf-Ekkehard Lönnig and Heinz-Albert Becker acknowledge that "it appears to be hard to even imagine a clearcut selective advantage for all the thousands of postulated intermediate steps in a gradual scenario... for the origin of the complex carnivorous plant structures examined above."

The development of any major new feature presents similar problems, and according to Lehigh University biochemist Michael Behe, who describes several spectacular examples in detail in *Darwin's Black Box* [Behe 1996], the world of microbiology is especially loaded with such examples of "irreducible complexity."

It seems that until the trigger hair, the door, and the pressurized chamber were all in place, and the ability to digest small animals, and to reset the trap to be able to catch more than one animal, had been developed, none of the individual components of this carnivorous trap would have been of any use. What is the selective advantage of an incomplete vacuum chamber? To the casual observer, it might seem that none of the components of this trap would have been of any use whatever until the trap was almost perfect, but of course a good Darwinist will imagine two or three far-fetched intermediate useful stages, and consider the problem solved. I believe you would need to find thousands of intermediate stages before this example of irreducible complexity has been reduced to steps small enough to be bridged by single random mutations—a lot of things have to happen behind the scenes and at the microscopic level before this trap could catch and digest animals. But I don't know how to prove this.

I am further sure that even if you could imagine a long chain of useful intermediate stages, each would present such a negligible selective advantage that nothing as clever as this carnivorous trap could ever be produced, but I can't prove that either. Finally, that natural selection seems even remotely plausible depends on the fact that while species are awaiting further improvements, their current complex structure is "locked in," and passed on perfectly through many generations (in fact, errors are constantly corrected and damage is constantly repaired). This phenomenon is observed, but inexplicable—I don't see any reason why all living organisms do not constantly decay into simpler components—as, in fact, they do as soon as they die.[3]

When you look at the individual steps in the development of life, Darwin's explanation is difficult to disprove, because some selective advantage can be imagined for almost anything. Like most other schemes designed to violate the second law, it is only when you look at the net result that it becomes obvious it won't work.

A November 2004 *National Geographic* article proclaims that the evidence is "overwhelming" that Darwin was right about evolution. Since there is no proof that natural selection has ever done anything more spectacular than cause bacteria to develop drug-resistant strains, where is the overwhelming evidence that justifies assigning to it an ability we do not attribute to any other natural force in the universe: the ability to create order out of disorder?

Three types of evidence are cited: first, the fact that species are so well suited to their environments is offered

3. Some Darwinists use computer programs, written by intelligent humans, which contain strings that simulate information in the DNA, and they run these programs on computers designed and maintained by intelligent humans. They introduce random errors into the strings, test the new strings for "fitness" in some way and discard the less fit strings, and claim the modest progress observed simulates evolution. If they want to see what unintelligent forces alone can accomplish, however, they should introduce random errors not only into the strings, but throughout the entire program, the compiler and operating system it uses, and the computer hardware. If you are trying to simulate how the accumulation of molecular accidents could produce complex organisms, why assume that only the DNA molecules are vulnerable to random damage? (This analogy was suggested by Gil Dodgen.)

as evidence that they have "adapted" to them. Of course, if they were not well-adapted, they would be extinct, and that would be offered as even stronger evidence against design. Second, they point to minor changes due to artificial selection, where intelligent humans select features already present in the gene pool, as evidence of what can be accomplished when natural forces select among genetic accidents. But, as always, the main evidence offered is the "evolutionary tree" of similarities connecting all species, fossil and living. These similarities were of course noticed long before Darwin (many animals have four legs, one head, two eyes and a tail!); all modern science has done is to show that the similarities go much deeper than those noticed by ancient man.

Although these similarities may, to our modern minds, suggest natural causes, they do not really tell us anything about what those causes might be. In fact, the fossil record does not even support the idea that new organs and new systems of organs arose gradually: new orders, classes and phyla consistently appear suddenly ("Gaps among known orders, classes and phyla are systematic and almost always large" [Simpson 1960]; see Section 3.2).

An analogy may be useful here. If some future paleontologist were to unearth two species of Volkswagens, he might find it plausible that one evolved gradually from the other. He might find the lack of gradual transitions between automobile families more problematic, for example, in the transition from mechanical to hydraulic brake systems, or from manual to automatic transmissions, or from steam engines to internal combustion engines; though if he thought about what gradual transitions would look like, he would understand why they didn't exist. He would be even more puzzled by the huge differences between the bicycle and motor vehicle phyla, or between the boat and airplane phyla. But heaven help us if he uncovers motorcycles and Hovercraft, the discovery of these "missing links" would be hailed in all our newspapers as final proof that all forms of transportation arose gradually from a common ancestor, without design.

Interestingly, although the similarities between species in the same branch of the evolutionary "tree" may suggest common descent, similarities (even genetic similarities) also frequently arise independently in distant branches, where they cannot be explained by common descent. This phenomenon, known as "convergence," suggests common design rather than common descent: the probability of similar designs arising independently through random processes is very small, but a designer could, of course, take a good design and apply it several times in different places, to unrelated species. For example, in their above-cited *Nature Encyclopedia of Life Sciences* article [Lönnig and Becker 2004] on Carnivorous Plants, Wolf-Ekkehard Lönnig and Heinz-Albert Becker note that "carnivory in plants must have arisen several times independently of each other... the pitchers might have arisen seven times separately, adhesive traps at least four times, snap traps two times and suction traps possibly also two times.... The independent origin of complex synorganized structures, which are often anatomically and physiologically very similar to each other, appears to be intrinsically unlikely to many authors so that they have tried to avoid the hypothesis of convergence as far as possible."

Since I am well aware that logic and evidence are powerless against the popular perception, nurtured by prestigious journals such as *National Geographic* and *Nature*, that no serious scientists harbor any doubts about Darwinism, I want to offer here a portion of a November 5, 1980 *New York Times News Service* report:

> Biology's understanding of how evolution works, which has long postulated a gradual process of Darwinian natural selection acting on genetic mutations, is undergoing its broadest and deepest revolution in nearly 50 years. At the heart of the revolution is something that might seem a paradox. Recent discoveries have only strengthened Darwin's epochal conclusion that all forms of life evolved from a common ancestor. Genetic analysis, for example, has shown that every organism is governed by the same genetic code controlling the same biochemical processes. At the same time, however, many studies suggest that the origin of species was not the way Darwin suggested.... Exactly how evolution happened is now a matter of great controversy among biologists. Although the debate has been under way for several years, it reached a crescendo last month, as some 150 scientists specializing in evolutionary studies met for four days in Chicago's Field Museum of Natural History to thrash out a variety of new hypotheses that are challenging older ideas. The meeting, which was closed to all but a few observers, included nearly all the leading evolutionists in paleontology, population genetics, taxonomy and related fields. No clear resolution of the controversies was in sight. This fact has often been exploited by religious fundamen-

talists who misunderstood it to suggest weakness in the fact of evolution rather than the perceived mechanism. Actually, it reflects significant progress toward a much deeper understanding of the history of life on Earth. At issue during the Chicago meeting was macroevolution, a term that is itself a matter of debate but which generally refers to the evolution of major differences, such as those separating species or larger classifications.... Darwin suggested that such major products of evolution were the results of very long periods of gradual natural selection, the mechanism that is widely accepted today as accounting for minor adaptations.... Darwin knew he was on shaky ground in extending natural selection to account for differences between major groups of organisms. The fossil record of his day showed no gradual transitions between such groups, but he suggested that further fossil discoveries would fill the missing links. 'The pattern that we were told to find for the last 120 years does not exist,' declared Niles Eldridge, a paleontologist from the American Museum of Natural History in New York. Eldridge reminded the meeting of what many fossil hunters have recognized as they trace the history of a species through successive layers of ancient sediments. Species simply appear at a given point in geologic time, persist largely unchanged for a few million years and then disappear. There are very few examples—some say none—of one species shading gradually into another.

5.4 Human Consciousness

For the layman, it is the last step in evolution that is the most difficult to explain. You may be able to convince him that natural selection can explain the appearance of complicated robots, who walk the Earth and write books and build computers, but you will have a harder time convincing him that a mechanical process such as natural selection could cause those robots to become conscious. Human consciousness is in fact the biggest problem of all for Darwinism, but it is hard to say anything "scientific" about consciousness, since we don't really know what it is, so it is also perhaps the least discussed.

Nevertheless, one way to appreciate the problem it poses for Darwinism or any other mechanical theory of evolution is to ask the question: is it possible that computers will someday experience consciousness? If you believe that a mechanical process such as natural selection could have produced consciousness once, it seems you *can't* say it could never happen again, and it might happen faster now, with intelligent designers helping this time. In fact, most Darwinists probably do believe it could and will happen—not because they have a higher opinion of computers than I do: everyone knows that in their most impressive displays of "intelligence," computers are just doing *exactly* what they are told to do, nothing more or less. They believe it will happen because they have a lower opinion of humans: they simply dumb down the definition of consciousness, and say that if a computer can pass a "Turing test," and fool a human at the keyboard in the next room into thinking he is chatting with another human, then the computer has to be considered to be intelligent, or conscious. With the right software, my laptop may already be able to pass a Turing test, and convince me that I am Instant Messaging another human. If I type in "My cat died last week" and the computer responds "I am saddened by the death of your cat," I'm pretty gullible, that might convince me that I'm talking to another human. But if I look at the software, I might find something like this:

```
if (verb == 'died')
  fprintf(1,'I am saddened by the death of your %s',noun)
end
```

I'm pretty sure there is more to human consciousness than this, and even if my laptop answers all my questions intelligently, I will still doubt there is "someone" inside my Intel processor who experiences the same consciousness that I do, and who is really saddened by the death of my cat, though I admit I can't prove that there isn't.

I really don't know how to argue with people who believe computers could be conscious. About all I can say is: what about typewriters? Typewriters also do exactly what they are told to do, and have produced some magnificent works of literature. Do you believe that typewriters can also be conscious?

And if you *don't* believe that intelligent engineers could ever cause computers to attain consciousness, how can you believe that random mutations could accomplish this?

5.5 Conclusions

Science has been so successful in explaining natural phenomena that the modern scientist is convinced that it can explain everything, and anything that doesn't fit into this model is simply ignored. It doesn't matter that there were no natural causes before Nature came into existence, so he cannot hope to ever explain the sudden creation of time, space, matter and energy and our universe in the big bang. It doesn't matter that quantum mechanics is based on a "principle of indeterminacy," that tells us that every "natural" phenomenon has a component that is forever beyond the ability of science to explain or predict, he still insists nothing is beyond the reach of his science. When he discovers that all of the basic constants of physics, such as the speed of light, the charge and mass of the electron, Planck's constant, etc., had to have almost exactly the values that they do have in order for any conceivable form of life to survive in our universe, he proposes the "anthropic principle" and says that there must be many other universes with the same laws, but random values for the basic constants, and one was bound to get the values right.

When you ask him how a mechanical process such as natural selection could cause human consciousness to arise out of inanimate matter, he doesn't understand what the problem is, and he talks about human evolution as if he were an outside observer, and never seems to wonder how he got inside one of the animals he is studying. And when you ask how the four fundamental forces of Nature could rearrange the basic particles of Nature into libraries full of encyclopedias, science texts and novels, and computers, connected to laser printers, LCDs and keyboards and the internet, he says, well, order can increase in an open system.

The development of life may have only violated one law of science, but that was the "supreme" law of Nature, and it has violated that in a most spectacular way. At least that is my opinion, but perhaps I am wrong.

Perhaps it only seems extremely improbable, but really isn't, that, under the right conditions, the influx of stellar energy into a planet could cause atoms to rearrange themselves into nuclear power plants and spaceships and computers. But one would think that at least this would be considered an open question, and those who argue that it really *is* extremely improbable, and thus contrary to the basic principle underlying the second law, would be given a measure of respect, and taken seriously by their colleagues, but we aren't.

5.6 Supplement: The Equations of Entropy Change

The following appears as Appendix D of my book "The Numerical Solution of Ordinary and Partial Differential Equations, second edition," John Wiley & Sons [Sewell 2005]. Reprinted with permission of John Wiley & Sons.

The partial differential equations that govern heat conduction and diffusion, derived in Section 2.0 and studied extensively in this book, have some relevance to an interesting philosophical question.

Consider the diffusion (conduction) of heat in a solid, R, with (absolute) temperature distribution $U(x, y, z, t)$. The first law of thermodynamics (conservation of energy) requires that

$$Q_t = -\nabla \bullet \mathbf{J} \tag{5.1}$$

where Q ($Q = c\rho U$) is the heat energy density and **J** is the heat flux vector. The second law requires that the flux be in a direction in which the temperature is decreasing, i.e.,

$$\mathbf{J} \bullet \nabla U \leq 0 \tag{5.2}$$

In fact, in an isotropic solid, **J** is in the direction of greatest decrease of temperature, that is, $\mathbf{J} = -K\nabla U$. Note that 5.2 simply says that heat flows from hot to cold regions—because the laws of probability favor a more uniform distribution of heat energy.

"Thermal entropy" is a quantity that is used to measure randomness in the distribution of heat. The rate of change of thermal entropy, S, is given by the usual definition as:

$$S_t = \iiint_R \frac{Q_t}{U} dV \tag{5.3}$$

Using 5.3 and the first law 5.1, we get:

$$S_t = \iiint_R \frac{-\mathbf{J} \bullet \nabla U}{U^2} dV - \iint_{\partial R} \frac{\mathbf{J} \bullet \mathbf{n}}{U} dA \tag{5.4}$$

where **n** is the outward unit normal on the boundary ∂R. From the second law (5.2), we see that the volume integral is nonnegative, and so

$$S_t \geq - \iint_{\partial R} \frac{\mathbf{J} \bullet \mathbf{n}}{U} dA \tag{5.5}$$

From 5.5 it follows that $S_t \geq 0$ in an isolated, closed, system, where there is no heat flux through the boundary ($\mathbf{J} \bullet \mathbf{n} = 0$). Hence, in a closed system, entropy can never decrease. Since thermal entropy measures randomness (disorder) in the distribution of heat, its opposite (negative) can be referred to as "thermal order," and we can say that the thermal order can never increase in a closed system.

Furthermore, there is really nothing special about "thermal" entropy. We can define another entropy, and another order, in exactly the same way, to measure randomness in the distribution of any other substance that diffuses, for example, we can let $U(x, y, z, t)$ represent the concentration of carbon diffusing in a solid (Q is just U now), and through an identical analysis show that the "carbon order" thus defined cannot increase in a closed system. It is a well-known prediction of the second law that, in a closed system, every type of order is unstable and must eventually decrease, as everything tends toward more probable states—not only will carbon and temperature distributions become more random (more uniform), but the performance of all electronic devices will deteriorate, not improve. Natural forces, such as corrosion, erosion, fire and explosions, do not create order, they destroy it. The second law is all about probability, it uses probability at the microscopic level to predict macroscopic change: the reason carbon distributes itself more and more uniformly in an insulated solid is, that is what the laws of probability predict, when diffusion alone is operative. The reason natural forces may turn a spaceship, or a TV set, or a computer into a pile of rubble but not vice-versa is also probability: of all the possible arrangements atoms could take, only a very small percentage could fly to the moon and back, or receive pictures and sound from the other side of the Earth, or add, subtract, multiply and divide real numbers with high accuracy.

The discovery that life on Earth developed through evolutionary "steps," coupled with the observation that mutations and natural selection—like other natural forces—can cause (minor) change, is widely accepted in the scientific world as proof that natural selection—alone among all natural forces—can create order out of disorder, and even design human brains, with human consciousness. Only the layman seems to see the problem with this logic. In a recent *Mathematical Intelligencer* article [Sewell 2000], after outlining the specific reasons why it is not reasonable to attribute the major steps in the development of life to natural selection, I asserted that the idea that the four fundamental forces of physics alone could rearrange the fundamental particles of Nature into spaceships, nuclear power plants, and computers, connected to laser printers, CRTs, keyboards and the internet, appears to violate the second law of thermodynamics in a spectacular way.

Anyone who has made such an argument is familiar with the standard reply: the Earth is an open system, and order can increase in an open system, as long as it is "compensated" somehow by a comparable or greater decrease outside the system. S. Angrist and L. Hepler in *Order and Chaos* [Angrist and Hepler 1967], write, "In a certain sense the development of civilization may appear contradictory to the second law.... Even though society can effect local reductions in entropy, the general and universal trend of entropy increase easily swamps the anomalous but important efforts of civilized man. Each localized, man-made or machine-made entropy decrease is accompanied by a greater increase in entropy of the surroundings, thereby maintaining the required increase in total entropy."

According to this reasoning, then, the second law does not prevent scrap metal from reorganizing itself into a computer in one room, as long as two computers in the next room are rusting into scrap metal—and the door is open. A closer look at equation 5.5, which holds not only for thermal entropy, but for the "entropy" associated with any other substance that diffuses, shows that this argument, which goes unchallenged in the scientific literature, is based on a misunderstanding of the second law. Equation 5.5 does not simply say that entropy cannot decrease in a closed system, it also says that in an open system, entropy cannot decrease faster than it is exported through the boundary, because the boundary integral there represents the rate that entropy is exported across the boundary: notice that the integrand is the outward heat flux divided by absolute temperature. (That this boundary integral represents the rate that entropy is exported seems to have been noticed by relatively few people [for example, Dixon 1975, p. 202], probably because the isotropic case is usually assumed and so the numerator is written as $-K\frac{\partial U}{\partial n}$, and in this form the conclusion is not as obvious.) Stated another way, the order in an open system cannot increase faster than it is imported through the boundary. According to 5.4, the thermal order in a system can decrease in two different ways, it can be converted to disorder (first integral term) or it can be exported through the boundary (boundary integral term). It can increase in only one way: by importation through the boundary. Similarly, the increase in "carbon order" in an open system cannot be greater than the carbon order imported through the boundary, and the increase in "chromium order" cannot be greater than the chromium order imported through the boundary, and so on.

The above analysis was published in my reply "Can ANYTHING Happen in an Open System?" [Sewell 2001] to critics of my original *Mathematical Intelligencer* article. In these simple examples, I assumed nothing but heat conduction or diffusion was going on, but for more general situations, I offered the tautology that *"if an increase in order is extremely improbable when a system is closed, it is still extremely improbable when the system is open, unless something is entering which makes it* not *extremely improbable."* The fact that order is disappearing in the next room does not make it any easier for computers to appear in our room—unless this order is disappearing *into* our room, and then only if it is a type of order that makes the appearance of computers not extremely improbable, for example, computers. Importing thermal order will make the temperature distribution less random, and importing carbon order will make the carbon distribution less random, but neither makes the formation of computers more probable. What happens in a closed system depends on the initial conditions; what happens in an open system depends on the boundary conditions as well.

As I wrote in [Sewell 2001], "order can increase in an open system, not because the laws of probability are suspended when the door is open, but simply because order may walk in through the door.... If we found evidence that DNA, auto parts, computer chips, and books entered through the Earth's atmosphere at some time in the past, then perhaps the appearance of humans, cars, computers, and encyclopedias on a previously barren planet could be explained without postulating a violation of the second law here (it would have been violated somewhere else!). But if all we see entering is radiation and meteorite fragments, it seems clear that what is entering through the boundary cannot explain the increase in order observed here."

6

My Failed Simulation

The following article appeared in the on-line ***Human Events*** *(www.humanevents.com/article.php?id=25030), February 15, 2008.*

The strongest argument for intelligent design is to clearly state the alternative view, which is that physics explains all of chemistry (probably true), chemistry explains all of biology, and biology completely explains the human mind; thus, physics alone explains the human mind. The following fictional thought experiment is designed to help those who dismiss intelligent design as unscientific, to think about what it is they really believe.

In a 2000 *Mathematical Intelligencer* article,[1] I speculated on what would happen if we constructed a gigantic computer model which starts with the initial conditions on Earth 4 billion years ago and tries to simulate the effects that the four known forces of physics (the gravitational and electromagnetic forces and the strong and

1. Chapter 3

weak nuclear forces) would have on every atom and every subatomic particle on our planet. If we ran such a simulation out to the present day, I asked, would it predict that the basic forces of Nature would reorganize the basic particles of Nature into libraries full of encyclopedias, science texts and novels, nuclear power plants, aircraft carriers with supersonic jets parked on deck, and computers connected to laser printers, CRTs and keyboards?

A friend read my article and said, computers have advanced a lot in the last seven years, I think we could actually try such a simulation on my new laptop now. So I wrote the program—in Fortran, naturally—and we tried it. It took several hours, and at the end of the simulation we dumped the final coordinates of all the particles into a rather large data file, then ran MATLAB to plot them. Some interesting things had happened, a few mountains and valleys and volcanos had formed, but no computers, no encyclopedias, and no cars or trucks. My friend said, let me see your program. After examining it, he exclaimed, no wonder, you treated the Earth as a closed system, order can't increase in a closed system. The Earth is an open system, you need to take into account the effect of the sun's energy. So I modified the boundary conditions to simulate the effect of the entering solar radiation, and reran it. This time some clouds and rivers had formed, but otherwise Earth still looked a lot like the other planets, and still no libraries or computers or airplanes.

My imaginary friend looked more carefully at the program, and said, good grief, you are using classical physics, you can't simulate the effects of the four forces without quantum mechanics. He explained that according to quantum mechanics, the exact effects of these forces on any particular particle are impossible to predict with certainty, the new laws only provide the probabilities. I said, you mean there is a supernatural force at work here? He said, well, technically, yes, if you define the supernatural to be that which is forever beyond the ability of science to predict or explain—British astronomer Sir Arthur Eddington said quantum mechanics "leaves us with no clear distinction between the natural and the supernatural." But there is no reason to doubt that this so-called "supernatural" effect is completely random, you can simulate it using a random number generator. So I completely re-wrote my simulator, I used an IMSL random number generator with a user-supplied probability distribution to simulate this randomness, and computed the required probability distributions by solving the Schrödinger equations with my own partial differential equation solver, PDE2D. Still no luck—no space ships, no TV sets, no encyclopedias, not even a cheap novel.

My friend looked at the new graphs and tried to mask his disappointment. Well, he said, of course the problem is you haven't taken into account the one natural force in the universe which *can* violate the second law of thermodynamics and create order out of disorder—natural selection. You mean there is a fifth force—why didn't you say so? Just give me the equations for this force and I will add it to my model. He said, I can't give you the equations, because it isn't actually a physical force, it doesn't actually move particles. So what does it do, I asked. He explained that one day a long time ago, by pure chance, a collection of atoms formed that was able to duplicate itself, and these complex collections of atoms were able to pass their complex structures on to their descendents generation after generation, even correcting errors. He went on to talk about how genetic accidents and survival of the fittest produced even more complex collections of atoms, and how something called "intelligence" allowed some of these collections of atoms to design computers and laser printers and the internet. But when he finished, I still didn't know how to incorporate natural selection—or intelligence—into my model, so I never did get the simulation to work. I decided the model was still missing a force or two—or a smarter random number generator.

7

How Evolution Will Be Taught Someday

The following article appeared in the on-line ***Human Events*** *(www.humanevents.com/article.php?id=26046), April 16, 2008, with original title: "Ben Stein's Expelled: A Little Background."*

With the release of Ben Stein's new movie *Expelled* on April 18, the question as to whether intelligent design is a scientific theory, or only religion in disguise, will be debated across the country as never before. Here is a little background, and a prediction as to how this controversy will play out in the coming years.

A November 5, 1980 *New York Times News Service* article, reporting on a meeting at the Chicago Field Museum of Natural History of "nearly all the leading evolutionists in paleontology, population genetics, taxonomy and related fields," begins:

> Biology's understanding of how evolution works, which has long postulated a gradual process

> of Darwinian natural selection acting on genetic mutations, is undergoing its broadest and deepest revolution in nearly 50 years. At the heart of the revolution is something that might seem a paradox. Recent discoveries have only strengthened Darwin's epochal conclusion that all forms of life evolved from a common ancestor.... At the same time, however, many studies suggest that the origin of species was not the way Darwin suggested... how evolution happened is now a matter of great controversy among biologists.[1]

These excerpts summarize nicely the main issues, which are really quite simple, in today's dispute between evolution and intelligent design (ID). On the one hand, there are many things about the development of life on Earth which suggest natural (unintelligent) causes: the long periods involved, the similarities between species, the many evolutionary dead ends, and so on. Furthermore, science has been so successful in explaining natural phenomena in other areas that many have come to believe the explanatory power of science has no limitations. On the other hand, Darwin's attempt to explain the origins of all the magnificent species in the living world in terms of the struggle for survival (easily the dumbest idea ever taken seriously by science, in my opinion) is rapidly losing support in the scientific community, as the true dimensions of the complexity of life are revealed by scientific research, especially at the microscopic level.[2] The major-

---

1. see Section 5.3 for the rest of this article.

2. A quick way to develop some appreciation for the complexity of the cell is to view the video, at www.signatureinthecell.com, associated with Stephen Meyer's new book [Meyer 2009]. For those with more time, I recommend the 600-page book.

ity still give lip-service to Darwinism, but only because all the alternative natural explanations which have been proposed are even more far-fetched.

This "paradox," as the *New York Times* article calls it, has left people searching for some middle ground. Many people feel silly attributing the origin of each species directly to God, yet understand that a completely unintelligent process could not possibly have produced what we see on Earth today.[3] For those who do not understand this, I recommend a little thought experiment, proposed in my February 15 *Human Events* essay "My Failed Simulation,"[4] which will help them think about what they have to believe, to not believe in intelligent design.

As Ben Stein's new movie documents, some good scientists now conclude that only intelligent causes can explain the incredible complexity of life. For example, biologist and genetic mutations expert Wolf-Ekkehard Lönnig of the Max Planck Institute for Breeding Research in Cologne, Germany, has written a very detailed, thoroughly researched, article (www.weloennig.de/Giraffe.pdf) "The Evolution of the Long-Necked Giraffe" which shows that nearly everything about the popular Darwinian story of how the giraffe got its long neck (including the idea that it happened gradually) is either false or unsubstantiated, and concludes, in Part II:

> ... the scientific data that are available to date on the question of the origin of the giraffe make a gradual development through mutation and

3. In an 1861 letter to Sir John Herschel, Charles Darwin wrote: "One cannot look at this Universe with all living productions and man without believing that all have been intelligently designed; yet when I look to each individual organism, I can see no evidence of this."

4. Chapter 6

> selection so extremely improbable that in any other area of life such improbability would force us to look for a feasible alternative. Yet biologists committed to a materialistic world view will simply not consider an alternative. For them, even the most stringent objections against the synthetic evolutionary theory are nothing but open problems that will be solved entirely within the boundaries of their theory. This is still true even when the trend is clearly running against them, that is, when the problems for the theory become greater and greater with new scientific data. This essential unfalsifiability, by the way, places today's evolutionary theory outside of science....

Dr. Lönnig, who has studied mutations for nearly 30 years, argues that intelligent design is the only possible explanation for the evolution of the modern-day giraffe from its short-necked, okapi-like ancestors. I am convinced that Dr. Lönnig is right, but he is ahead of his time. Will we ever see the day when intelligent design is taught as a scientific explanation for the origin of species in high school and university biology classrooms? Perhaps, but probably not in my lifetime.

A much more likely result of this paradox is that in the not-too-distant future, biology texts will refer to evolution as an amazing, mysterious "natural" process, which scientists do not now understand, but hope to understand some day. Natural selection may then be mentioned only as a historical footnote, as a very simplistic early attempt to explain the workings of this natural process.

But for most ID proponents, this will be a quite satisfactory outcome, certainly a huge improvement over the current sad state of affairs, where Darwin's natural selection is the only scientific theory around which enjoys widespread legal protection from scientific criticism in the classroom. The Discovery Institute, a leading promoter of ID as a scientific theory, does *not* (contrary to common belief) support the teaching of intelligent design in science classrooms, they only hope that biology instructors will be allowed to "teach the scientific controversy" over Darwinism.

Perhaps after a few generations in which biology texts confidently predict that future discoveries will uncover the mechanism of evolution, eventually some will begin to recognize the obvious, that there is no possible explanation without design. Until then, I will be happy with texts which simply acknowledge that the idea that the survival of the fittest can turn bacteria into giraffes, and cause human consciousness to arise out of inanimate matter, is doubted by some scientists.

8

The Supernatural Element in Nature

8.1 Axioms and Evidence

In his 1888 book *Evolution* [Le Conte 1888] Joseph Le Conte, professor of Geology and Natural History at the University of California, writes:

> Intermediate links may be wanting now, but they must, of course, have existed once—i.e., in previous geological times, and therefore ought to be found fossil. In distribution in space or geographically, organic kinds may be marked off by hard-and-fast lines but, if their derivative origin be true, in their distribution in time or geologically, there ought to be many examples of insensible shadings between them. In fact, if we only had all the extinct forms, the organic kingdom, taken as a whole and throughout all time, ought to consist not of species at all, but simply of individual forms, shading in-

> sensibly into each other.... But this is not the fact. On the contrary, the law of distribution in time is apparently similar in this respect to the law of distribution in space, already given. As in the case of contiguous geographical faunas, the change is apparently by substitution of one species for another, and not by transmutation of one species into another. So also in successive geological faunas, the change seems rather by substitution than by transmutation. In both cases species seem to come in suddenly, with all their specific characters perfect, remain substantially unchanged as long as they last, and then die out and are replaced by others. Certainly this looks much like immutability of specific forms, and supernaturalism of specific origin.... The reason for this, given by Darwin and other evolutionists, is the extremely fragmentary character of the geological record.... While it is true that there are many and wide gaps in the record... yet there are some cases where the record is not only continuous for hundreds of feet in thickness, but the abundance of life was very great, and the conditions necessary for preservation exceptionally good... and yet, although the species change greatly, and perhaps many times, in passing from the lowest to the highest strata, we do not usually, it must be acknowledged, find the gradual transitions we would naturally expect if the changes were effected by gradual transformations.

Le Conte also acknowledges that natural selection cannot explain the appearance of new, irreducibly complex, features ("novelties"):

> ... neither can it [natural selection] explain the first steps of advance toward usefulness. An organ must be already useful before natural selection can take hold of it to improve on it.

After acknowledging that the only direct evidence, the fossil record, does not support the idea of gradual change, and that the only theory ever taken seriously as to the causes of these changes cannot explain anything new, Le Conte nevertheless concludes:

> We are confident that evolution is absolutely certain—not evolution as a special theory —Lamarckian, Darwinian, Spencerian... but evolution as a law of derivation of forms from previous forms. In this sense it is not only certain, it is *axiomatic*.... The origins of new phenomena are often obscure, even inexplicable, but we never think to doubt that they have a natural cause; for so to doubt is to doubt the validity of reason, and the rational constitution of Nature.

Le Conte illustrates the optimism which prevailed in science in the late 19th century. Science had made such progress explaining previously mysterious phenomena that there was no reason to believe, it was felt, that any of the secrets of Nature, even the secrets of life itself, would long endure the assault of scientific investigation. In Le Conte's day, nearly all scientists held the view that everything that happens in our world is completely determined by the laws of Nature, that the only limits to our ability to understand what has happened, and predict what will happen in the future, are practical limits on the extent of our knowledge.

Le Conte's axiom that science can explain everything continued as a fundamental pillar of philosophy throughout the 20th century. Olan Hyndman, in *The Origin of Life and the Evolution of Living Things*, [Hyndman 1952], calls Darwinism "the most irrational and illogical explanation of natural phenomena extant." Yet he says "I have one strong faith, that scientific phenomena are invariable... any exception is as unthinkable as to maintain that thunderbolts are tossed at us by a manlike god named Zeus," and so he goes on to develop an alternative—and even more illogical—theory (Lamarckian, basically) of the causes of evolution. Jean Rostand [Rostand 1956], quoted in previous chapters, says "However obscure the causes of evolution appear to me to be, I do not doubt for a moment that they are entirely natural." Hans Gaffron [Gaffron 1960], in a paper presented at the 1959 University of Chicago Centennial Congress *Evolution after Darwin*, presents a theory on the origin of life, but admits, "no shred of evidence, no single fact whatever, forces us to believe in it. What exists is only the scientists' wish not to admit a discontinuity in Nature and not to assume a creative act forever beyond comprehension."

A recent (November 10, 2008) article in *News at Princeton* (www.princeton.edu/main/news/archive) entitled "Evolution's New Wrinkle: Proteins with Cruise Control Provide New Perspective," reports on research by four Princeton scientists, published in a *Physical Review Letters* article:

> The experiments, conducted in Princeton's Frick Laboratory, focused on a complex of proteins located in the mitochondria, the powerhouses

> of the cell.... Chakrabarti and Rabitz analyzed these observations of the proteins' behavior from a mathematical standpoint, concluding that it would be statistically impossible for this self-correcting behavior to be random, and demonstrating that the observed result is precisely that predicted by the equations of control theory.... The authors sought to identify the underlying cause for this self-correcting behavior in the observed protein chains. Standard evolutionary theory offered no clues.... Chakrabarti said, 'Control theory offers a direct explanation for an otherwise perplexing observation and indicates that evolution is operating according to principles that every engineer knows.' The scientists do not know how the cellular machinery guiding this process may have originated, but they emphatically said it does not buttress the case for intelligent design.

No explanation whatever is offered for why the authors reject the conclusion to which their experiments and observations seem to point. None is needed, because everyone understands the reason: Le Conte's axiom.

8.2 The Advent of Quantum Mechanics

Surprisingly, less than 40 years after his book appeared, Le Conte's axiom was shattered by the discoveries of quantum mechanics, which introduced, quite literally, a "supernatural" element into science.

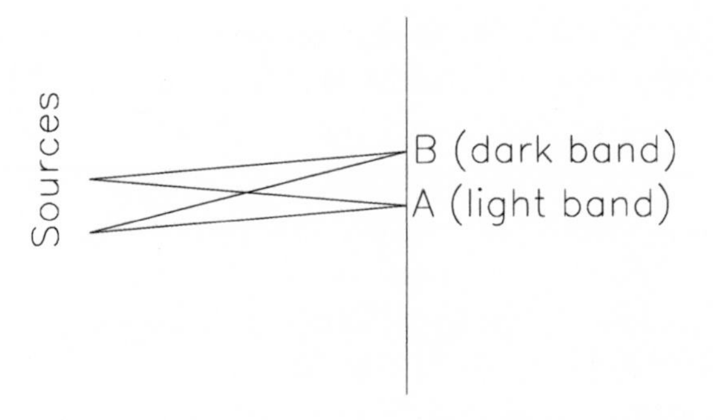

Figure 8.1. Wave Diffraction

To understand the background for the discoveries of quantum mechanics, let us start with a classic diffraction experiment. Suppose two wave sources in phase and of the same wavelength, λ, are placed a small distance apart as shown in Figure 8.1. We can imagine these to be sound waves, for example. At a point on a wall, A, chosen to be equally distant from each source, the waves from the two sources will arrive in phase, and reinforce each other. However, at a point B, chosen to be exactly $\lambda/2$ further from one source than the other, the waves will arrive one half cycle out of phase, and cancel each other at all times. We will also observe this cancellation at the points whose distances from the two sources differ by 1.5, 2.5, 3.5,... wavelengths, and so as we move up the wall we will encounter alternating points of reinforcement and cancellation. Experiments with light diffraction had been carried out, in which light from a distant source, of a single wavelength, is passed through two narrow slits on a plate perpendicular to the direction to the source. Since the two slits are equally distant from the source, the light should hit the two slits in phase, and the two slits can thus be considered to be separate light sources in phase with each other. Where the light from these two "sources" hits a wall, a diffraction pattern with alternating light and dark bands will be observed. If one slit is covered up, the dark bands go away!

It is easy to see why, at the beginning of the 20th century, it was unanimously agreed that light must consist of waves. If light consists of particles, it is hard to see how light from one source could cancel the light from another source!

Some new experiments, however, seemed to be inconsistent with the wave theory of light. In the photoelectric effect, for example, it was observed that when a metal plate was illuminated, the energy delivered by the light caused some electrons to be stripped from their host atoms and ejected from the plate. Since an electron must reach a certain threshold energy level before it can escape the metal, experimenters were surprised to find that even when light of extremely low intensity was aimed at the plate, a few electrons were immediately able to absorb enough energy to be ejected. If light were propagated through waves we would expect the light energy to be spread more evenly over the metal, and at very low intensities we would expect to have to wait a while before *any* electron could absorb enough energy to escape. When the intensity of the light was increased, another unexpected result was observed. The number of electrons emitted increased with the intensity, but the energies of the individual emitted electrons were unchanged. The ejected electrons, it seemed, had received a packet, or "quantum," of energy whose magnitude was independent of the light intensity; increasing the intensity seemed only to increase the number of such packets available.

A particle theory of light would explain these results: even at very low light intensities, a few electrons (those hit by the light "particles") would be immediately ejected, and increasing the intensity (the number of light particles bombarding the metal) would cause more electrons to be knocked out, but the energy of an individual ejected electron would depend only on the energy of the light particle which struck it, not on the number of light particles. Further experimentation showed that while the energy of an individual ejected electron did not vary with the intensity of the light, it did change with color, increasing as the wavelength of the light was decreased.

For a while, light had to be considered to have a dual nature, since some experiments (such as diffraction) could only be explained using the wave theory, while others (such as the photoelectric effect) could only be explained using the particle theory. The spectroscope, a tool used by astronomers, separates out the different wavelengths of light by bending them through different angles. It was designed using the wave theory of light and it should not work, according to the particle theory. The Geiger counter, on the other hand, is designed to count individual "particles" of electromagnetic radiation, and light is electromagnetic radiation.

In the early 1920's, the two opposing views of light were reconciled by the following theory: Light consists of particles (photons), but there is a wave associated with each photon, whose intensity at a given point gives the *probability* of finding a photon at that point. In other words, light consists of particles whose motions are guided by probability waves.

In 1924 French physicist Louis de Broglie further suggested that this dual wave/particle nature was characteristic not only of electromagnetic waves, such as light, but of all "particles." He concluded that any particle of momentum p is guided by a probability wave of wavelength $\lambda = h/p$, where h is called Planck's constant. This would explain why, in the photoelectric experiment, the electrons knocked out by the lower wavelength light came off with a higher energy: the lower wavelength photons have a greater momentum. Spectacular confirmation of de Broglie's conclusion came in 1927, when electron diffraction was first observed, by Davisson and Germer at Bell Telephone Laboratories. The electron's particle nature was undisputed: we find 1,2,3... electrons in an atom; we never find the electronic charge or mass in other than integral multiples. Yet electrons were observed to diffract—a phenomenon unique to wave motion, involving cancellation—when passed through a metal crystal. Because electrons typically pack a much greater momentum than photons, and thus their associated wavelengths are much smaller, the electron diffraction pattern is only observable when the spacing between slits is very small. That is why electron diffraction was first observed using the tiny spaces between atoms in a metal crystal as "slits." Other atomic particles such as neutrons have since been made to exhibit the diffraction characteristic of waves as well.

The governing equation of the new quantum physics is the Schrödinger equation, which can be used to calculate the "probability distribution" of particles. For example, Figure 8.2 [taken from Fitzgerald and Sewell 2000] shows the probability distribution associated with the second lowest energy level, for an electron in the vicinity of two protons, as calculated by my partial differential equation solver, PDE2D (www.vni.com/pde2d). Note that there is no attempt to say exactly where the electron is at any given time (until it is directly observed), we can only say where it "probably" is.

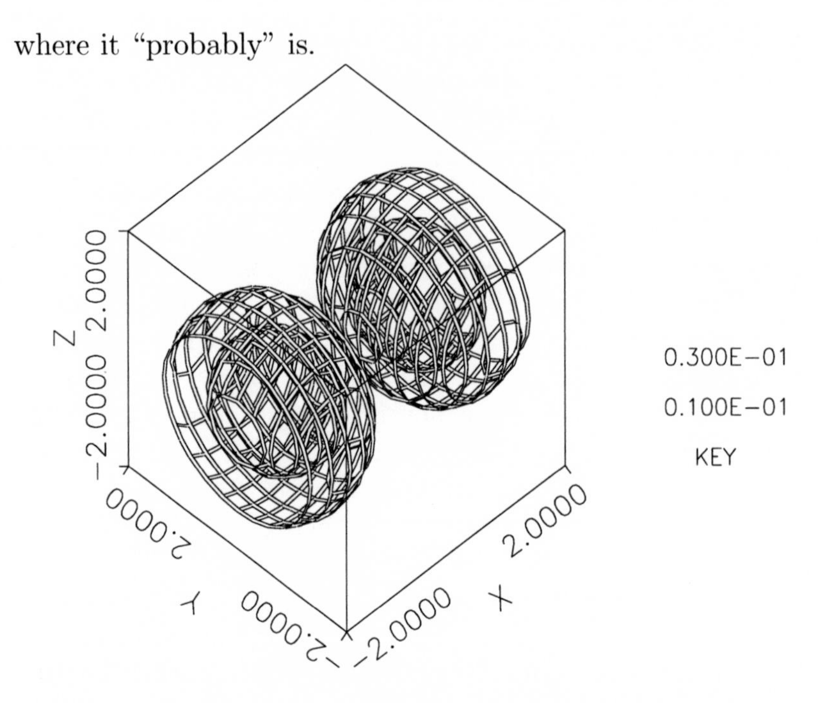

Figure 8.2. Electron Probability Density Near Two Protons
(levels coded by color in original)

To fully appreciate why science was forced, to the dismay of many, to drag "probability" into the picture, let us go back and repeat the two-slit diffraction problem of Figure 8.1, only this time let us imagine a beam of electrons rather than a beam of light, and let us replace the wall with a photographic plate. (We need to also imagine that the spacing between slits is extremely small, since these are electrons.)

Let us set the intensity of the electron beam at such a low level that we can assume that only one electron at a time passes through the diffraction apparatus. Each electron which is not stopped by the first plate will pass through one of the two slits and hit the photographic plate at a particular point, marking its impact with a dark spot. After these dark spots begin to accumulate, however, we begin to observe the familiar wave diffraction pattern of alternating light and dark bands on the film. The individual electrons impact the film at specific spots, yet the collection of impact marks conforms to the diffraction pattern expected for a wave whose wavelength is given by the de Broglie formula. In other words, a particular electron may hit the film almost anywhere, but when a large number of electrons pass through the slits, the result is highly predictable.

Suppose we repeat the experiment, only this time instead of leaving both slits open long enough for N electrons to pass through, we block the top slit and leave the bottom slit open until N/2 electrons have passed through it; then we block the bottom slit and let another N/2 electrons pass through the top slit. Surely the results will be the same as in the first experiment. How could it possibly matter whether we allow the electrons to alternate randomly between slits, or force the first batch of electrons to pass through the bottom slit and the second batch to pass through the top slit? But it does matter: in the first experiment we would get a diffraction pattern, while in the second we would get only a more or less uniformly exposed film. Incredibly, the behavior of an electron passing through one slit seems to be affected by whether or not it *could have* passed through the other! We can explain these results only if, when both slits are open, we think of each individual electron as a probability wave, passing through both slits—and yet each electron strikes the film as a particle! In other words, until it is actually observed, we must think of the position of the electron as *inherently* ill-defined, specified only by a probability density function; when it is finally observed (when it hits the film) it has a very definite position.

8.3 Philosophical Implications

The introduction of "probability" into physics has enormous philosophical implications. For the first time, science had to face the fact that no matter how well we prepare for any experiment, no matter how much data we accumulate, we cannot predict with certainty the outcome of the experiment. British astronomer Sir Authur Eddington, in his classic work *The Nature of the Physical World* [Eddington 1929], says that according to quantum theory, "the future is a combination of the causal influences of the past together with unpredictable elements—unpredictable not merely because it is impracticable to obtain the data of prediction, but because no data connected causally with our experience exist."

Einstein objected to quantum mechanics with its introduction of chance and the "uncertainty principle" into science, saying "God does not play dice," but the quantum theory has been so successful in explaining scientific phenomena that it is now universally accepted.

If there are a billion electrons in an electron beam, we can predict with high accuracy and high confidence the pattern they will make as they hit the target, but if we look at a particular electron in the beam we can predict where it will hit with much less accuracy. If we have a billion atoms of a radioactive substance whose half-life is 10 years, we can be very confident that almost exactly a half billion will decay within 10 years, but if we try to predict when a particular atom will decay, all we can do is guess. And it doesn't matter how much we learn about that particular electron or radioactive atom, or its neighbors, we will never be able to predict with certainty what the electron or atom will do. For it is not the practical constraints of our experiment, but the theory itself, that limits our predictive powers.

One of the philosophical implications of the "uncertainty principle" introduced by quantum mechanics is that the idea—so contrary already to our intuition—that all human actions are strictly determined (in a complicated way) by external influences, is shown once and for all to be wrong. For even the individual particles which make up the brain have a "free will" of their own; even their behavior is not strictly predictable. Eddington says [Eddington 1929], "It is meaningless to say that the behavior of a conscious brain is precisely the same as that of a mechanical brain, if the behavior of a mechanical brain is left undetermined." Further, he states that with the advent of quantum mechanics, "science thereby withdraws its moral opposition to freewill."

It could be added that science must also withdraw its moral opposition to religion, for if we define the "supernatural" to be that which is forever beyond the ability of science to predict or explain, then there is, quite literally, a "supernatural" element to all "natural" phenomena. Eddington says that quantum mechanics "leaves us with no clear distinction between the natural and the supernatural."

When we say that the result of a coin toss, for example, is determined by "chance," we really mean that it is determined by factors too complicated to predict in practice, but we assume that if we knew the initial conditions and forces with sufficient accuracy we could predict whether it would land heads or tails. But with quantum mechanics, when we talk about "chance," we mean something very different, we do not mean a factor too complicated to predict in practice, but rather a factor which is *inherently* impossible to predict. Although science can still be used to predict macroscopic phenomena with probabilities approaching certainty, it can predict microscopic phenomena with less confidence.

We have already seen in Chapter 1 that the discovery of the big bang means that atheists can no longer complain that those who believe that "in the beginning God created the heavens and the Earth" are profaning science with supernatural speculation. Since there were no natural causes before Nature came into existence suddenly a few billion years ago, they are now just as guilty; now everyone must speculate about the supernatural forces which created our universe, the debate is now only about whether those forces were intelligent or unintelligent. Now the same can be said about the origin of life, or the origin of species: the atheist can no longer criticize proponents of intelligent design for staining the purity of science by trying to introduce supernatural causes into the picture (this criticism is still used, of course, and quite liberally, but it is no longer logically valid). Now it must be accepted by everyone—everyone who is aware of quantum mechanics, at least—that there is a supernatural component to all natural phenomena, the question is again only whether this supernatural component is intelligent or unintelligent. And while it is difficult to see any clear and compelling evidence of intelligent design in many "natural" phenomena, when we look at the origin and development of life, the evidence is overwhelming. Even if we accept the Darwinist's claim—wholly unsupported by the evidence—that species developed very gradually, the question is still there: was the supernatural element involved intelligent or truly "random," as Darwin believed?

In summary, those who claim that science has eliminated the supernatural from Nature have a view of science that has been out of date for 80 years. When we try to reduce all of reality to matter in motion, we find quite a surprise: there at the bottom, controlling the motion of matter, is the remarkable Schrödinger equation of quantum mechanics, which tells us that science is an entertaining and useful tool to help us understand our world, but it does not have all the answers, and *never will.*

9

The Scientific Theory of Intelligent Design

In a December 21, 2005 on-line *American Spectator* article, Jay Homnick wrote:

> It is not enough to say that design is a more likely scenario to explain a world full of well-designed things. Once you allow the intellect to consider that an elaborate organism with trillions of microscopic interactive components can be an accident... you have essentially 'lost your mind.'

Before Darwin, nearly everyone, in every corner of the world, believed in some type of "intelligent design" (the majority still do), for good reasons. Since the publication of *Origin of Species*, science has discovered that living things are far more complex and clever than Darwin could have ever imagined, and Darwin's explanation for this complexity has become less and less plausible, so the reasons for believing in intelligent design have only *increased* in the last 150 years. Even atheist Richard

Dawkins wrote [Dawkins 1996] that "biology is the study of complicated things that give the *appearance* of having been designed." So how did it happen that a majority of our intellectuals lost their minds?

I think I can explain. When one becomes a scientist, one learns that science can now explain so many previously inexplicable phenomena that one comes to believe that nothing can escape the explanatory power of our science. When one becomes a biologist, or a paleontologist, one discovers many things about the origin of species, such as the long periods involved and the evidence for common descent, that give the impression of natural causes. When one studies history, one may become overwhelmed by the misery and confusion of the human condition, and wonder, why is it so hard to see evidence of the hand of God in human history?

But notably absent from any list of reasons why intellectuals reject intelligent design (ID) is any direct scientific evidence that natural selection of random mutations or any other unintelligent process can actually do intelligent things, like design plants or animals. The argument "we have found natural explanations for many other previously unexplained phenomena" is powerful, but not definitive: there are numerous examples in the history of science of ideas that worked well for a long time, then quit working when applied in new situations. The arguments "this just looks like natural causes" and "why is the world God created sometimes so cruel?" are even more persuasive—there *are* many things in the history of life that leave a strong impression of natural causes, and I certainly do understand why the pain and evil that exist in this world cause many to doubt that it is designed—but these arguments are obviously also not conclusive (see the Epilogue). Darwinists have discovered that they can simply line up a series of similar fossils in a museum and count on this as being mistaken as conclusive evidence for "natural" causes. A series of fossils is not a scientific argument against design, it is a theological argument against design; it does not tell us *anything* about the causes of the changes. The argument is basically, "this doesn't look like the way God would have created things," an argument frequently used by Darwin in *Origin of Species.* (In fact, it does look a lot like the way we create things, through testing and improvements; see Section 3.2.) However strong may be the philosophical, psychological and religious reasons why many of our greatest minds reject it, the argument for intelligent design is still crystal clear to the unindoctrinated: unintelligent forces cannot design eyes, ears, hearts and brains. This argument *is* definitive, and from a purely logical point of view, much more powerful than all the others together. No matter how many other mysteries of Nature may yield to scientific investigation, and no matter how much evidence for common descent we may find, Jay Homnick is still right. Once you allow yourself to seriously consider the possibility that the human body and the human brain could be entirely the products of unintelligent forces, you have lost your mind.

Nevertheless, Le Conte's axiom (Section 8.1) that everything must have a natural explanation has become the foundation of all of modern thought—and indeed it has proven to be a very useful and productive axiom. Even many people who believe in God accept Le Conte's axiom. "Theistic evolutionists" argue that God created the universe and its laws, and that these laws are sufficient to explain everything we see today. I have no philosophical or theological problem with such a view: the laws God created are very cleverly designed, and they alone probably are sufficient to explain all of chemistry, geology, astronomy and atmospheric science, for example, so it is not surprising that many would insist that it must be possible to explain all of biology using these laws as well. The problem I have with this view is logical: the known laws of physics are indeed very cleverly designed, and may explain everything that has happened on *other* planets, but they are obviously *not* clever enough to explain all of biology. The atheistic evolutionist has decided *a priori* that there can be no design in Nature; the theistic evolutionist has decided *a priori* that there can be design only in the original laws of Nature. ID proponents argue that we should look at the evidence before deciding where there is design.

Le Conte's axiom is useful, and philosophically attractive to many. But to abandon belief in an intelligent designer simply because of such an axiom is circular reasoning of the worst sort, yet this is exactly the sort of reasoning on which the whole of modern philosophy is based.

I have been working on this book for about 30 years now, and some of the material in the first few chapters was written in the late 70s or early 80s; some of this material appeared in a "Postscript" of a 1985 Springer Verlag book (see Chapter 4) and in a couple of self-published booklets, because in those days it was impossible to publish anything critical of evolution in the scientific journals. Then, the only other people questioning Darwin's explanation for the development of life were the so-called "creationists," who were right on the main issue, but who had a much broader agenda, they were trying to make the case for a literal interpretation of Genesis, and debates over evolution often became debates over the historical accuracy of the Biblical account of Noah and the flood. The modern resurgence of "intelligent design," which perhaps began with the 1996 publication of Michael Behe's landmark work, *Darwin's Black Box* [Behe 1996] and which is led by the Discovery Institute's Center for Science and Culture (CSC) (www.discovery.org/csc), is quite different from yesterday's creationist movement. ID supporters claim, quite correctly, that one can deduce the existence of an intelligent designer from the evidence all around us, particularly in biology, but do not attempt to go any further than that, because that is as far as the scientific evidence alone can take us. In fact, many ID proponents will not even identify this designer with God, because, they say, the scientific evidence does not tell us anything about the designer. For all we know, for example, he could have been a "more evolved" visitor from another planet, as Richard Dawkins speculated in the movie *Expelled.* This is true; however, the fine tuning of the laws of physics (Section 2.1) obviously cannot be credited to Dawkins' alien, and it seems reasonable to assume it was the same designer, and the designer of the laws of Nature is "God" by definition—by my definition, at least. Naturally, the scientific evidence does not tell us if this is the God of the Bible, or of the Koran, or an "unknown" God.

Many critics of ID today still try to label ID as "creationism," because it was so much easier to discredit the old creationists—all you had to do was produce evidence for an old Earth, or for common descent, then you didn't have to deal with their main point. Others avoid the real issue by simply dismissing ID as "not science." Even some scientists who have concluded that life could not have developed without design, still argue that such a conclusion is not science! I have noticed that in debates, when ID proponents cite examples of irreducible complexity, such as the carnivorous plants (Section 5.3), or spectacular examples from microbiology such as the bacterial flagellum (which has all the parts of an outboard motor) or other features described in *Darwin's Black Box* [Behe 1996], Darwinists rarely even speculate on how these features *might* have evolved through small improvements; they simply dismiss ID as "not science." The attitude of the majority of scientists today still seems to be that ID must be discarded *a priori*, before even evaluating the supporting evidence, because ID is not science. We recall from Chapter 1 that the idea that time could have a beginning was also rejected *a priori* by many scientists—Walther Nernst said it "contradicted the very foundations of science." The big bang theory nevertheless eventually won acceptance due to the "pressure of empirical data."

There is a question I would like to see posed, in such a debate, to one of these critics who dismiss ID as not science: "suppose we *did* discover some biological feature that was irreducibly complex, to your satisfaction—such a spectacular example of irreducible complexity that you and every reasonable person would agree that it could not have evolved through small improvements—would the design hypothesis *then* be justified?" Of course, there are thousands of features in every living cell which any unbiased observer would recognize as irreducibly complex, but suppose we found one that was still more spectacular by far. If he answers, yes, we just haven't found any such thing yet, then all the constantly-repeated philosophical objections that "ID is not science" immediately fall, because he has admitted that design is a legitimate, even if currently unjustified, scientific hypothesis. If the answer is, no, then everyone will finally understand that, as W. E. Lönnig has stated, today's evolutionary theory is completely unfalsifiable—there is no amount of evidence that will change these people's minds.[1]

Perhaps nothing illustrates just how "unfalsifiable" today's evolutionary theory really is than the reaction (or rather, lack of reaction) to the "front-loading" being discovered by modern science in the genes of primitive animals. Consider, for example, this report from a recent article in *Science* [Pennisi 2008]:

> Trichoplax adhaerens barely qualifies as an animal. About 1 mm long and covered with cilia, this flat marine organism lacks a stomach, muscles, nerves, and gonads, even a head... yet this animal's genome looks surprisingly like ours, says Daniel Rokhsar, an evolutionary biologist at the University of California, Berkeley. Its 98 million DNA base pairs include many of the genes responsible for guiding the development of other animals' complex shapes and organs, he and his colleagues report in the 21 August issue of *Nature*.... Adds Casey Dunn, an evolutionary biologist at Brown University, 'It is now completely clear that genomic complexity was present very early on' in animal evolution....

1. Darwin himself, in *Origin of Species*, acknowledged that the existence of irreducibly complex features would falsify Darwinism: "If it could be demonstrated that any complex organ existed which could not possibly have been formed by numerous, successive, slight modifications, my theory would absolutely break down." But, clearly, the gradual development of useful features through useless stages would be as hard to explain without intelligence as their sudden appearance, so not only Darwinism, but any theory without design, would "absolutely break down."

> 'Many genes viewed as having particular functions in bilaterians or mammals turn out to have a much deeper evolutionary history than expected, raising questions about why they evolved,' says Douglas Erwin, an evolutionary biologist at the Smithsonian National Museum of Natural History in Washington.

Front-loading is completely fatal to Darwinism: there is no *possible* selective advantage for the possession of genes for traits which would not evolve until millions of years later! Yet for today's evolutionary biologists, such discoveries only "raise questions about why they evolved." They seem completely incapable of drawing the obvious conclusion, that processes incapable of planning ahead—incapable of *design*—could not possibly cause genes to appear long before the traits they control exist.[2]

I was a visiting assistant professor at Purdue University in 1978-79, when I replied to a letter in the Purdue student newspaper (the *Exponent*) which compared those who doubt Darwin to "flat earthers." My response anticipates today's debate over whether intelligent design is science or not:

> Last year I surveyed the literature on evolution in the biology library of Oak Ridge National Laboratory and found Olan Hyndman's *The Origin of Life and the Evolution of Living Things* in which he calls the neo-Darwinian theory of random mutation and natural selection 'the most irrational and illogical explanation of natural phenomenon extant' and proposes an

2. But programmers can and occasionally do include code which is not used by the current version of their program, in anticipation of a need for it in some future version.

> alternative theory; Rene Dubos' *The Torch of Life* in which he says '[The neo-Darwinian theory's] real strength is that however implausible it may appear to its opponents they do not have a more plausible one to offer in its place'; and Jean Rostand's[3] *A Biologist's View* in which he says that the variations which made up evolution must have been 'creative and not random.' Rostand, who elsewhere has called the neo-Darwinian theory a 'fairy tale for adults,' attributes this creativeness to the genes themselves, and says 'quite a number of biologists do, in fact, fall back on these hypothetical variations to explain the major steps of evolution.' I was not, however, able to find any books which suggested that this creativeness originated outside the chromosomes—these are restricted to theological libraries, because they deal with religion and not science, and their authors are compared to flat earthers in *Exponent* letters.

It is argued that ID is not science because its proponents are not sure, and cannot agree among themselves, exactly when, where or how design came into play in evolution—was design involved in the creation of every new species or, for example, front-loaded into the genes of the new animal phyla which appeared suddenly during the Cambrian explosion 500 million years ago? Were new species "special creations," or actual descendents of previous species, the products of "creative" mutations?[4] These are legitimate objections, and certainly ID is a sci-

3. Rostand was "one of the leading European biologists," according to the book cover, and the author of more than thirty scientific books.

4. "Most mutations that built the great structures of life must have been nonrandom," concludes Michael Behe [Behe 2007].

ence in its infancy, but by looking at the larger view, we can be sure there was design involved in the history of life, even if we do not yet know when, where or how. It is possible that we will never know. But even if we concede that ID is not science, and thus should not be taught in the science classroom,[5] that does not justify teaching bad science as established fact. Why can't we simply tell students the truth, that we know nothing about the origin of life, and hardly anything about the causes of evolution? Why can't students be told that we know virtually nothing about the origin of species (or at least of orders, classes and phyla), and allowed to draw their own conclusions as to whether the forces that created eyes, ears, hearts and brains were intelligent or unintelligent? Of course, if students are told that the causes of evolution are unknown, or even controversial, most will revert back to the default, common sense explanation, design. This explains why Darwinists feel the need for the constant intimidation of opponents and suppression of opposing viewpoints that we see today in academia, and why they insist on telling students there is no controversy over causes among scientists, when there obviously is.

It is commonly believed that Darwinism is based on good science, and that those who oppose it simply do not like its philosophical and religious implications. The truth is exactly the opposite: Darwinism is based on very bad science, and those who support it do so primarily because they do not like what they see as the alternative. If you really examine the reasons individual scientists support Darwinism, I believe you will find

5. The Discovery Institute CSC opposes any effort to require the teaching of ID in science classes, it only promotes teaching the "strengths and weakness" of neo-Darwinian evolutionary theory, see www.discovery.org/a/3164

in almost every case that they are philosophical ("this just looks like natural causes"; "we have found natural explanations for so many other previously inexplicable phenomena") and theological (problems with the Bible, disillusion with organized religion, the problem of pain). For example, Darwin wrote, "I can indeed hardly see how anyone ought to wish Christianity to be true; for if so the plain language of the text seems to show that the men who do not believe, and this would include my father, brother and almost all my best friends, will be everlastingly punished."[6] The fact that Darwin's explanation is still the scientific orthodoxy, and that his opponents are still labeled kooks, is due not to the reasonableness of Darwinism, or to any shortage of evidence for design in Nature, but to the theological and philosophical problems its supporters have with what they see as the only alternative. If a plausible materialistic theory were to be developed, I have no doubt that Darwinism would be quickly dropped by the scientific community. No such alternative theory is on the horizon, however.

Although there is no scientific support for the idea that natural selection of random mutations, or any other unintelligent cause, can explain the complexity of life, some of the philosophical and theological reasons scientists have for rejecting the design alternative are pretty powerful, and I think I understand most of them well. I also wonder why God would create things the way He did, why we can only see Him so indirectly. I also have

---

6. Darwin is apparently referring to passages like John 3:18, "He who does not believe is condemned," which are sometimes interpreted to mean that all non-Christians are "condemned." If I thought the Christian God were that unfair, I would share Darwin's view of Christianity, but that John did not mean this as a condemnation of all non-Christians is clear from the following verse: "... and this is the condemnation, that light has come into the world, and men preferred the darkness, because their deeds were evil."

problems with the Bible, I am also embarrassed by many things religious leaders do and say, and I have especially struggled with the problem of pain: why is there so much misery in the world God created?

Because I believe the strongest objections to ID are philosophical rather than scientific, I feel this book would be incomplete if it did not make any attempt to address any of these philosophical problems. Thus in the Epilogue I will try to address what I believe is the strongest objection to ID, the problem of pain.

Epilogue: Is God Really Good?

The following is a slightly-modified version of an article which appeared in the Sri Aurobindo International Centre of Education (Pondicherry, India) on-line journal ***AntiMatters*** *(www.anti-matters.org) February 29, 2008.*

E.1 Is God Really Good?

Why do bad things happen to good people? This is the question which Rabbi Harold Kushner, in his highly-acclaimed 1981 book [Kushner 1981] called "the only question which really matters" to his congregation. It is a question which has been asked by philosophers and ordinary human beings throughout the ages; if not the most-asked question, certainly the most passionately-asked. It was certainly the first question that occurred to me in 1987 when I was told that my beloved wife Melissa, 34 years old and the mother of our two small children (Chris and Kevin), had cancer of the nose and sinuses, and in 1990 when we discovered that the cancer had recurred. The suffering she bravely endured during those years was beyond description, from the aggressive chemotherapy treatments, each of which required hospitalization for severe nausea and other side effects, from the radiation therapy, and from three major surgeries. Before the last surgery, during which they would remove her left eye and half of her teeth, she said, well, many people would be happy to have one eye. The cancer recurred two months after this surgery and I was terribly depressed for many years after her death. Since I am a pretty logical person, it never occurred to me to ask "does God really exist?", but I certainly wondered, "is God really good?"

Melissa Wehmann Sewell (1953-1991) with Chris

I think most people who claim not to believe in God, say this not because of any shortage of evidence for design in Nature, but because it is sometimes so hard to see evidence that God cares about us, and they prefer not to believe in God at all, than to believe in a God who doesn't care.

A wonderful little article in *UpReach* [Nov-Dec 1984] by Batsell Barrett Baxter, entitled "Is God Really Good?", contains some insights into the "problem of pain," as C. S. Lewis calls it [Lewis 1962], which I have found very useful. I will follow Baxter's outline in presenting my own thoughts on this question, and I would like to begin with his conclusion: "As I have faced the tragedy of evil in our world and have tried to analyze its origin, I have come to the conclusion that it was an inevitable accompaniment of our greatest blessings and benefits." In his outline, Baxter lists some examples of blessings which have, as inevitable consequences, unhappy side effects. None of these points is likely to make suffering in its severest forms any easier to accept, and we may be left wondering whether these blessings are really worth the high cost. But I believe they do at least point us in the right direction.

E.2 The Regularity of Natural Law

The laws of Nature which God has made work together to create a magnificent world, of oceans and forests, mountains and rivers, jungles and waterfalls, planets and stars, animals and plants. The basic laws of physics are cleverly designed to create conditions on Earth suitable for human life and human development. Gravity prevents us and our belongings from floating off into space; water makes our crops grow; the fact that certain materials are combustible makes it possible to cook our food and stay warm in winter. Yet gravity, water and fire are responsible for many tragedies, such as airplane crashes, drownings and chemical plant explosions. Tragedies such as floods and automobile accidents are the results of laws of physics which, viewed as a whole, are magnificently designed and normally work for our benefit. Nearly everything in Nature which is harmful to man has also a benevolent side, or is the result of a good thing gone bad. Even pain and fear themselves have useful purposes: pain warns us that something in our body needs attention, and without fear, we would all die young doing foolish and dangerous things.

But why won't God protect us from the bad side effects of Nature? Why doesn't He overrule the laws of Nature when they work against us? Why is He so "silent" during our most difficult and heart-breaking moments? First of all, if we assume He has complete control over Nature, we are assuming much more than we have a right to assume. It does not necessarily follow that, because something is designed, it can never break down. We design cars, and yet they don't always function as designed. When our car breaks down, we don't conclude that the designer planned for it to break down, nor do we conclude that it had no designer; when the human body breaks down, we should not jump to the conclusion that God planned the illness, nor should we conclude that the body had no designer.

That we were designed by a fantastically intelligent super intellect is a conclusion which is easily drawn from the evidence all around us. To jump from this to the conclusion that this creator can control *everything* is quite a leap. In fact, I find it easy to draw the opposite conclusion from the evidence, that this creator *cannot*, or at least does not, control everything. And even if we assume He has complete control over Nature it is hard to see how He could satisfy everyone. Your crops are dry so you pray for rain—but I am planning a picnic. It seems more fair to let Nature take its course and hope we learn to adapt.

In any case, what would life be like if the laws of Nature were not reliable? What if God could and did stand by to intervene on our behalf every time we needed Him? We would then be spared all of life's disappointments and failures, and life would certainly be less dangerous, but let us think about what life would be like in a world where nothing could ever go wrong.

I enjoy climbing mountains—small ones. I recently climbed an 8,700 foot peak in the Guadalupe Mountains National Park and was hot and exhausted, but elated, when I finished the climb. Later I heard a rumor that the Park Service was considering building a cable car line to the top, and I was horrified. Why was I horrified—that would make it much easier for me to reach the peak? Because, of course, the pleasure I derived from climbing that peak did not come simply from reaching the top—it came from knowing that I had faced a challenge and overcome it. Since riding in a cable car requires no effort, it is impossible to fail to reach the top, and thus taking a cable car to the peak brings no sense of accomplishment. Even if I went up the hard way again, just knowing that I could have ridden the cable car would cheapen my accomplishment.

When we think about it, we see in other situations that achieving a goal brings satisfaction only if effort is required, and only if the danger of failure is real. And if the danger of failure is real, sometimes we will fail.

When we prepare for an athletic contest, we know what the rules are and we plan our strategy accordingly. We work hard, physically and mentally, to get ready for the game. If we win, we are happy knowing that we played fairly, followed the rules, and achieved our goal. Of course we may lose, but what satisfaction would we derive from winning a game whose rules are constantly being modified to make sure we win? It is impossible to experience the thrill of victory without risking the agony of defeat. How many fans would attend a football game whose participants are just actors, acting out a script which calls for the home team to win? We would all rather go to a real game and risk defeat.

Life is a real game, not a rigged one. We know what the rules are, and we plan accordingly. We know that the laws of Nature and of life do not bend at our every wish, and it is precisely this knowledge which makes our achievements meaningful. If the rules of Nature were constantly modified to make sure we achieved our goals—whether they involve proving Fermat's Last Theorem, getting a book published, finding a cure for Alzheimer's disease, earning a college degree, or making a small business work—we would derive no satisfaction from reaching those goals. If the rules were even occasionally bent, we would soon realize that the game was rigged, and just knowing that the rules were flexible would cheapen all our accomplishments. Perhaps I should say, "if we were aware that the rules were being bent," because I do believe that God has at times intervened in human and natural history, and I would like to believe He still does so on occasions, but we are at least left with the strong impression that the rules are inflexible.

If great works of art, music, literature, or science could be realized without great effort, and if success in such endeavors were guaranteed, the works of Michelangelo, Mozart, Shakespeare and Newton would not earn much admiration. If it were possible to realize great engineering projects without careful study, clever planning and hard work, or without running any risk of failure, mankind would feel no satisfaction in having built the Panama Canal or having sent a man to the moon. And if the dangers Columbus faced in sailing into uncharted waters were not real, we would not honor him as a brave explorer. Scientific and technological progress are only made through great effort and careful study, and not every scientist or inventor is fortunate enough to leave his mark, but anyone who thinks God would be doing us a favor by dropping a book from the sky with all the answers in it does not understand human nature very well—that would take all the fun out of discovery. If the laws of Nature were more easily circumvented, life would certainly be less frustrating and less dangerous, but also less challenging and less interesting.

Many of the tragedies, failures and disappointments which afflict mankind are inevitable consequences of laws of Nature and of life which, viewed as a whole, are magnificently designed and normally work for our benefit. And it is because we know these laws are reliable, and do not bend to satisfy our needs, that our greatest achievements have meaning.

E.3 The Freedom of Man's Will

I believe, however, that the unhappiness in this world attributable to "acts of God" (more properly called "acts of Nature") is small compared to the unhappiness which we inflict on each other. Reform the human spirit and you have solved the problems of drug addiction, drunk driving, war, broken marriages, child abuse, neglect of the elderly, crime, corruption and racial hatred. I suspect that many (not all, of course) of the problems which we generally blame on circumstances beyond our control are really caused by, or aggravated by, man—or at least could be prevented if we spent as much time trying to solve the world's problems as we spend in hedonistic pursuits.

God has given us, on this Earth, the tools and resources necessary to construct, not a paradise, but something not too far from it. I am convinced that the majority of the things which make us most unhappy are the direct or indirect result of the sins and errors of people. Often, unfortunately, it is not the guilty person who suffers.

But our evil actions are also the inevitable result of one of our highest blessings—our free will. C. S. Lewis, in *Mere Christianity* [Lewis 1943], says, "Free will, though it makes evil possible, is also the only thing that makes possible any love or goodness or joy worth having.... Someone once asked me, 'Why did God make a creature of such rotten stuff that it went wrong?' The better stuff a creature is made of—the cleverer and stronger and freer it is—then the better it will be if it goes right, but also the worse it will be if it goes wrong."

Why do a husband and wife decide to have a child? A toy doll requires much less work, and does not throw a temper tantrum every time you make him take a bath or go to bed. A stuffed animal would be much less likely to mark on the walls with a crayon, or gripe about a meal which took hours to prepare. But most parents feel that the bad experiences in raising a real child are a price worth paying for the rewards—the hand-made valentine he brings home from school, and the "I love you" she whispers as she gives her mother and father a good night kiss. They recognize that the same free will which makes a child more difficult to take care of than a stuffed animal also makes him more interesting. This must be the way our Creator feels about us. The freedom which God has given to us results, as an inevitable consequence, in many headaches for Him and for ourselves, but it is precisely this freedom which makes us more interesting than the other animals. God must feel that the headaches are a price worth paying: He has not taken back our free will, despite all the evil we have done. Why are there concentration camps in the world that God created? How could the Christian church sponsor the Crusades and the Inquisition? These terribly hard questions have a simple answer: because God gave us all a free will.

Jesus told a parable about "wheat and tares," which seems to teach that the weeds of sin and sorrow cannot be eliminated from the Earth without destroying the soil of human freedom from which the wheat of joy and goodness also springs. It is impossible to rid the world of the sorrow caused by pride, selfishness and hatred without eliminating the free will which is also the source of all the unselfishness and love that there is in the world. Thought itself is an expression of our free will, and to say that God ought to prevent us from doing evil is to request that our ability to think be withdrawn. If we ask God to take back the free will which He has given us, we might as well ask Him to turn us into rocks.

If we base our view of mankind on what we see on the television news, we may feel that good and evil are greatly out of balance today; that there is much more pain than joy in the world, and much more evil than goodness. It is true that the amount of pain which exists in our world is overwhelming, but so is the amount of happiness. And if we look more closely at the lives of those around us, we will see that the soil of human freedom still produces wheat as well as weeds. The dark night of Nazi Germany gave birth to the heroism of Dietrich Bonhoeffer, Corrie ten Boom and many others. The well-known play "The Effect of Gamma Rays on Man-in-the-Moon Marigolds" is about two sisters raised by a bitter mother who suffocates ambition and discourages education. One sister ends up following the path to destruction taken by her mother; the other refuses to be trapped by her environment, and rises above it. It may seem at times that our world is choking on the weeds of pain and evil, but if we look closely we will see that wheat is still growing here.

Again we conclude that evil and unhappiness are the inevitable by-products of one of our most priceless blessings—our human free will.

E.4 The Interdependence of Human Lives

Since it is our human free will which makes our relationships with others meaningful, his third point is closely related to the second, but Baxter nevertheless considers this point to be important enough to merit separate consideration.

Much of an individual's suffering is the direct or indirect result of the actions or misfortunes of others. Much of our deepest pain is the result of loneliness caused by the loss of the love or the life of a loved one, or of the strain of a bad relationship. How much suffering could be avoided if only we were "islands, apart to ourselves." Then at least we would suffer only for our own actions, and feel only our own misfortunes. The interdependence of human life is certainly the cause of much unhappiness.

Yet here again, this sorrow is the inevitable result of one of our greatest blessings. The pain which comes from separation is in proportion to the joy which the relationship provided. Friction between friends is a source of grief, but friendship is the source of much joy. Bad marriages and strained parent-child relationships are responsible for much of the unhappiness in the modern world, but none of the other joys of life compare to those which can be experienced in a happy home. Although real love is terribly hard to find, anyone who has experienced it—as I did for a few short years—will agree that the male-female relationship is truly a masterpiece of design, when it works as it was intended to work.

As Baxter writes, "I am convinced that our greatest blessings come from the love which we give to others and the love which we receive from others. Without this interconnectedness, life would be barren and largely meaningless. The avoidance of all contact with other human beings might save us some suffering, but it would cost us the greatest joys and pleasures of life."

E.5 The Value of Imperfect Conditions

We have thus far looked at suffering as a by-product of our blessings and not a blessing in itself. And certainly it is difficult to see anything good in suffering in its severest forms.

Nevertheless, we cannot help but notice that some suffering is necessary to enable us to experience life in its fullest, and to bring us to a closer relationship with God. Often it is through suffering that we experience the love of God, and discover the love of family and friends, in deepest measure. The man who has never experienced any setbacks or disappointments invariably is a shallow person, while one who has suffered is usually better able to empathize with others. Some of the closest and most beautiful relationships occur between people who have suffered similar sorrows.

It has been argued that most of the great works of literature, art and music were the products of suffering. One whose life has led him to expect continued comfort and ease is not likely to make the sacrifices necessary to produce anything of great and lasting value.

Of course, beyond a certain point pain and suffering lose their positive value. Even so, the human spirit is amazing for its resilience, and many people have found cause to thank God even in seemingly unbearable situations. While serving time in a Nazi concentration camp for giving sanctuary to Jews, Betsie ten Boom [ten Boom 1971] told her sister, "We must tell people what we have learned here. We must tell them that there is no pit so deep that God is not deeper still. They will listen to us, Corrie, because we have been here." C. S. Lewis concludes his essay on *The Problem of Pain* [Lewis 1962] by saying "Pain provides an opportunity for heroism; the opportunity is seized with surprising frequency." We might add that not only the person who suffers, but also those who minister to his needs, are provided with opportunities for growth and development.

As Baxter put it: "The problems, imperfections and challenges which our world contains give us opportunities for growth and development which would otherwise be impossible."

E.6 Conclusions

In *Brave New World* [Huxley 1932], Aldous Huxley paints a picture of a futuristic Utopian society which has succeeded, through totalitarian controls on human behavior and drugs designed to stimulate pleasant emotions and to repress undesirable ones, in banishing all traces of pain and unpleasantness. There remains one "savage" who has not adapted to the new civilization, however, and his refusal to take his pills results in the following interchange between "Savage" and his "civilized" interrogators:

"We prefer to do things comfortably," said the Controller.

"But I don't want comfort, I want God, I want poetry, I want real danger, I want freedom, I want goodness, I want sin."

"In fact," said Mustophe Mond, "you're claiming the right to be unhappy."

"Alright then," said the Savage defiantly, "I'm claiming the right to be unhappy."

If God designed this world as a tourist resort where man could rest in comfort and ease, it is certainly a dismal failure. But I believe, with Savage, that man was created for greater things. That is why, I believe, this world presents us with such an inexhaustible array of puzzles in mathematics, physics, astronomy, biology and philosophy to challenge and entertain us, and provides us with so many opportunities for creativity and achievement in music, literature, art, athletics, business, technology and other pursuits; and why there are always new worlds to discover, from the mountains and jungles of South America and the flora and fauna of Africa, to the era of dinosaurs and the surface of Mars, and the astonishing world of microbiology.

Why does God remain backstage, hidden from view, working behind the scenes while we act out our parts in the human drama? This question has lurked just below the surface throughout much of this book, and now perhaps we finally have an answer. If He were to walk out onto the stage, and take on a more direct and visible role, I suppose He could clean up our act, and rid the world of pain and evil—and doubt. But our human drama would be turned into a divine puppet show, and it would cost us some of our greatest blessings: the regularity of natural law which makes our achievements meaningful; the free will which makes us more interesting than robots; the love which we can receive from and give to others; and even the opportunity to grow and develop through suffering. I must confess that I still often wonder if the blessings are worth the terrible price, but God has chosen to create a world where both good and evil can flourish, rather than one where neither can exist. He has chosen to create a world of greatness and infamy, of love and hatred, and of joy and pain, rather than one of mindless robots or unfeeling puppets.

Batsell Barrett Baxter, who was dying of cancer as he wrote these words, concludes: "When one sees all of life and understands the reasons behind life's suffering, I believe he will agree with the judgment which God Himself declared in the Genesis story of creation: 'And God saw everything that He had made, and behold it was *very* good.'"

References

Angrist, S. and L.Hepler (1967) *Order and Chaos*, Basic Books.

Behe, Michael (1996) *Darwin's Black Box*, Free Press.

Behe, Michael (2007) *The Edge of Evolution*, Free Press.

Daubenmire, R.F. (1947) *Plants and Environment*, John Wiley & Sons.

Davies, Paul (1980) *Other Worlds*, Simon and Schuster.

Davis, Tom (2001) "The Credibility of Evolution," *The Mathematical Intelligencer* 23, number 3, 4-5.

Dawkins, Richard (1996) *The Blind Watchmaker*, W. W. Norton.

Dembski, William (2006) *The Design Inference*, Cambridge Studies in Probability, Induction and Decision Theory.

Dixon, J. (1975) *Thermodynamics I: An Introduction to Energy*, Prentice-Hall.

Eddington, Arthur (1929) *The Nature of the Physical World*, McMillan.

Fitzgerald, Rosa and Granville Sewell (2000), "Solving Problems in Computational Physics Using a General-

Purpose PDE Solver," *Computer Physics Communications* 124, 132-138.

Ford, Kenneth (1968) *Basic Physics*, Blaisdell Publishing Co.

Gaffron, Hans (1960) "The History of Life," in Volume I of *Evolution after Darwin*, University of Chicago Press.

Gonzalez, Guillermo and Jay Richards (2004) *The Privileged Planet*, Regnery Publishing.

Guth, Alan (1998) *The Inflationary Universe*, Basic Books.

Harrison, Edward (1981) *Cosmology*, Cambridge University Press.

Hawking, Steven (1988) *A Brief History of Time—From the Big Bang to Black Holes*, Bantam Books.

Heeren, Fred (1995) *Show Me God*, Searchlight Publications.

Huxley, Aldous (1932) *Brave New World*, Bantam Books.

Hyndman, Olan (1952) *The Origin of Life and the Evolution of Living Things*, Philosophical Library.

Jastrow, Robert (1978) *God and the Astronomers*, W.W.Norton.

Kushner, Harold (1981) *When Bad Things Happen to Good People*, Schocken Books.

Le Conte, Joseph (1888) *Evolution*, D.Appleton and Company.

Leggett, A.J. (1987) *The Problems of Physics*, Oxford University Press.

Lewis, C. S. (1943) *Mere Christianity*, MacMillan.

Lewis, C. S. (1962) *The Problem of Pain*, MacMillan.

Lönnig, Wolf-Ekkehard and Heinz-Albert Becker (2004) "Carnivorous Plants," in *Nature Encyclopedia of Life Sciences*, Nature Publishing Group, Wiley Interscience. (http://mrw.interscience.wiley.com/emrw/9780470015902/els/article/a0003818/current/abstract)

Meyer, Stephen (2009) *Signature in the Cell*, HarperOne.

Pecker, Jean-Claude and Jayant Narlikar (2006) *Current Issues in Cosmology*, Cambridge University Press.

Pennisi, Elizabeth (2008) "Genomics: 'Simple' Animal's Genome Proves Unexpectedly Complex," *Science* 321, number 5892, 1028-1029.

Rosenhouse, Jason (2001) "How Anti-Evolutionists Abuse Mathematics," *The Mathematical Intelligencer* 23, number 4, 3-8.

Rostand, Jean (1956) *A Biologist's View*, Wm. Heinemann Ltd.

Sewell, Granville (1985) *Analysis of a Finite Element Method: PDE/PROTRAN*, Springer Verlag.

Sewell, Granville (2000) "A Mathematician's View of Evolution," *The Mathematical Intelligencer* 22, number 4, 5-7.

Sewell, Granville (2001) "Can ANYTHING Happen in an Open System?," *The Mathematical Intelligencer* 23, number 4, 8-10.

Sewell, Granville (2005) *The Numerical Solution of Ordinary and Partial Differential Equations, second edition*, John Wiley & Sons.

Simpson, George Gaylord (1960) "The History of Life," in Volume I of *Evolution after Darwin*, University of Chicago Press.

Strauss, Walter (2008) *Partial Differential Equations, An Introduction, second edition*, John Wiley & Sons.

ten Boom, Corrie (1971) *The Hiding Place*, Chosen Books.

Urone, Paul Peter (2001) *College Physics*, Brooks/Cole.

Varghese, Roy, editor (1984) *The Intellectuals Speak Out About God*, Regnery Gateway.

von Weizsäcker, C. F. (1964) *The Relevance of Science*, Harper & Row.

Ward, Peter and Donald Brownlee (2003) *Rare Earth: Why Complex Life is Uncommon in the Universe*, Springer Verlag.

Wells, Jonathan (2006) *Darwinism and Intelligent Design*, Regnery Publishing, Inc.

Weinberg, Steven (1977) *The First Three Minutes*, Basic Books.

Zel'Dovich, Y.B. and I.D.Novikov (1983) *The Structure and Evolution of the Universe*, University of Chicago Press.

Index

Breinigsville, PA USA
24 January 2010
231219BV00002B/4/P